W9-DEU-204

ECOSYSTEM PLANNING IN FLORIDA

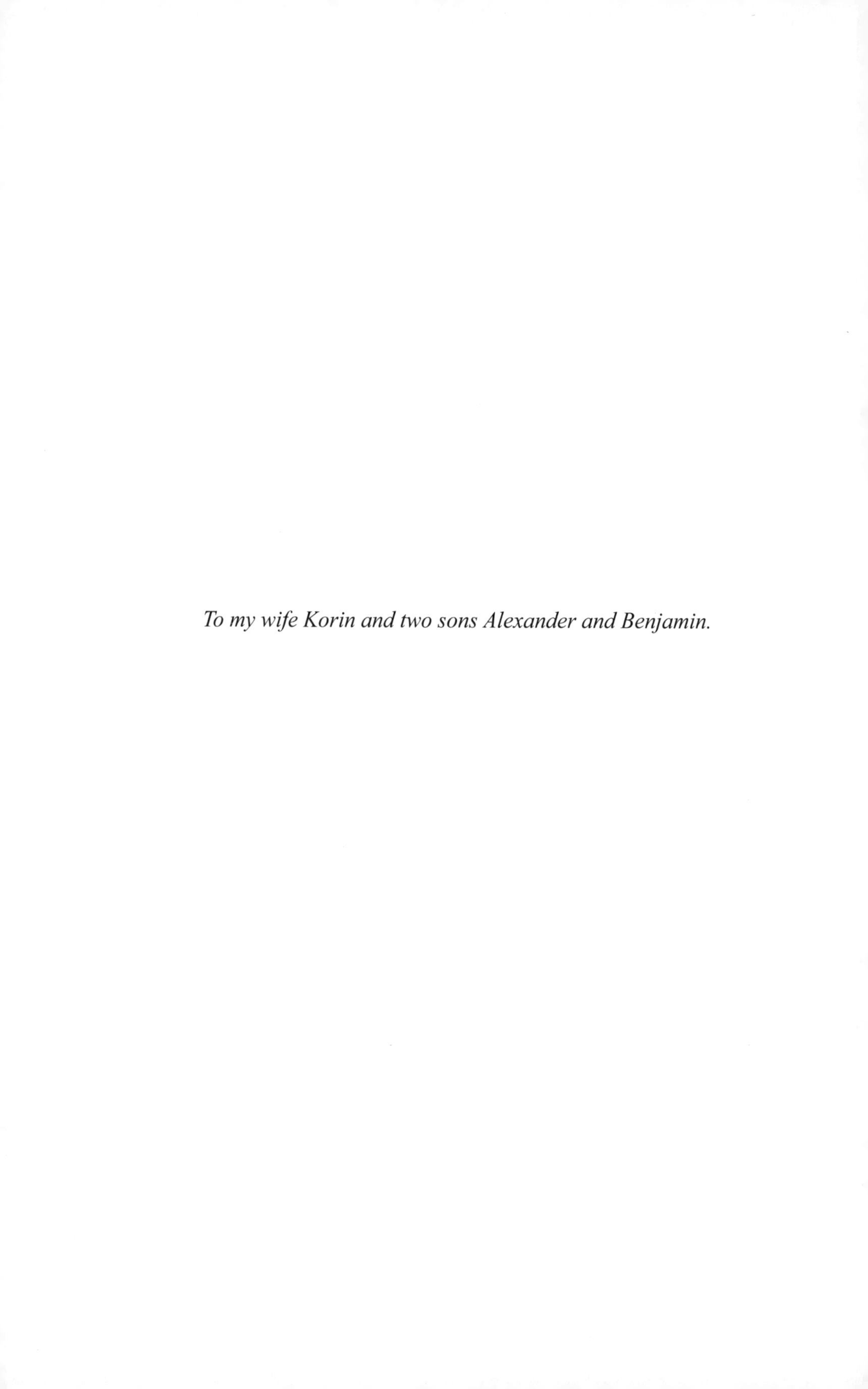

To my wife Korin and two sons Alexander and Benjamin.

Ecosystem Planning in Florida

Solving Regional Problems through Local Decision-making

SAMUEL DAVID BRODY

Texas A&M University, USA

ASHGATE

Published by
Ashgate Publishing Limited
Gower House
Croft Road
Aldershot
Hampshire GU11 3HR
England

Ashgate Publishing Company
Suite 420
101 Cherry Street
Burlington, VT 05401-4405
USA

Ashgate website: http://www.ashgate.com

British Library Cataloguing in Publication Data
Brody, Samuel David
Ecosystem planning in Florida : solving regional problems through local decision-making
1. Ecosystem management - Florida - Planning 2. Ecosystem management - Florida - Decision making
I. Title
333.7'8216'09759

Library of Congress Cataloging in Publication Data
Brody, Samuel David.
Ecosystem planning in Florida : solving regional problems through local decision-making / by Samuel David Brody.
p. cm.
Includes index.
1. Ecosystem management--Florida--Planning. 2. Ecosystem management--Florida--Decision making. I. Title.

QH76.5.F6B76 2008
333.95'1609759--dc22

2008002623

ISBN 978-0-7546-7249-4

Printed and bound in Great Britain by MPG Books Ltd, Bodmin, Cornwall.

Contents

List of Figures

List of Tables

Notes on Author

Samuel D. Brody is an Associate Professor of Environmental Planning in the Department of Landscape Architecture and Urban Planning at Texas A&M University. He is the Director of the Environmental Planning and Sustainability Research Unit, Co-Director of the Center for Texas Beaches and Shores, and a faculty fellow in the Hazard Reduction and Recovery Center. Dr. Brody's research focuses on environmental planning, spatial analysis, environmental dispute resolution, climate change policy, and natural hazards mitigation. Dr. Brody teaches graduate courses in environmental planning, sustainable development, and dispute resolution. He has also worked in both the public and private sectors to help local coastal communities to draft land use and environmental plans.

Preface

This book culminates over five years of research on ecosystem-based planning in Florida. It is meant to provide an evidenced-based examination of the topic that can inform both practitioners and academics on how to manage ecological systems from a local-level perspective. Such an approach stands in contrast to previous work, which has been largely argumentative, descriptive, and based on isolated case studies. The motivation for this book thus comes from a strong need for an empirically-driven, quantitative assessment of what works and what does not in terms of incorporating the needs of ecological systems into local planning frameworks.

The book is broken down into two major components: concepts and evidence-based case studies that put the concepts to work. Readers who are primarily interested in the applications of the research or empirical evidence can proceed directly to the case studies. Those who desire more of a theoretical grounding can concentrate on the chapters preceding the case studies. All readers should cover the integrated set of recommendations presented in Chapter 12. These recommendations offer the most insight on how to make use of the study's findings.

The content and conclusions of this book are useful to a diverse set of audiences, including: academics and those affiliated with universities who teach and research in the areas of environmental studies, environmental planning, and natural resource management; practicing local and regional planners, policy makers, and nongovernmental organizations with an interest in guiding development while at the same time protecting the structure and function of ecological systems; consultants engaged in projects involving local and regional environmental planning issues; and finally, advanced undergraduate and graduate level students taking courses in departments such as planning, ecology, wildlife and fisheries, parks and recreation, geography, and public policy.

Acknowledgements

This book is based on research supported in part by the US National Science Foundation Grant No. CMS-0346673 to the Texas A&M University, the US Environmental Protection Agency STAR Grant No. U-915600, the Lincoln Land Institute, and the Texas A&M University Sustainable Coastal Margins Program. The findings and opinions reported are those of the author and are not necessarily endorsed by the funding organizations or those who provided assistance with various aspects of the study.

I would like to thank my research assistants Wesley E. Highfield, Jennifer Yust-Dyke, Sara Thornton, and Virginia Carrasco. Without their many contributions and hard work this research could have never been accomplished. I also thank my colleagues Michael Lindell and Walter G. Peacock at Texas A&M University for their constant encouragement and support. Appreciation is due to the Department of Planning at Florida Atlantic University, which housed me as a visiting scholar during which many of the pages of this book were written. Thanks also go to my mentors at the University of North Carolina at Chapel Hill, Raymond Burby, David Godschalk, and Phillip Berke for their invaluable training and for paving the way with their previous work.

Finally, and most importantly, I want to thank my family who provided unfailing support, patience, and encouragement throughout my career. Principally, my wife Korin, without whose enduring love and belief in my abilities I could not have completed this work. Also, our two sons Alexander and Benjamin for always providing needed perspective. Thanks go to my mother, Linda H. Brody, for opening my eyes to new possibilities and teaching me through her own life the meaning of the phrase "fall down seven times, get up eight." She is an inspiration to me and so many others. Gratitude also goes to my father Paul S. Brody for his daily patience, guidance, and friendship. Final appreciation is also due to Janet and Marty Wilk for their support throughout this project.

List of Abbreviations

DCA	Department of Community Affairs
DEP	Department of Environmental Protection
EMA	Ecosystem Management Area
EUP	Eastern Upper Peninsula of Michigan Ecosystem Management Project
FLU	Future Land Use
FPL	Florida Power and Light
GIS	Geographic Information Systems
HUC	Hydrological Unit Code
NEP	National Estuary Program
NGO	Non-government Organization
NIPF	Non-industrial Private Forests
UBG	Urban Growth Boundary
US	United States
USGS	United States Geological Survey
WMD	Water Management District

Chapter 1

Ecosystem Planning at the Local Level: An Introduction

The United States (US) boasts one of the most comprehensive and longstanding environmental policy frameworks in the world. Despite its myriad of environmental programs, regulations, and permitting processes, the Country has been unable to halt the decline of its critical natural resources. Increasing development and consumption of natural systems have resulted in adverse impacts to water quality, loss of habitat, and the overall reduction of biological diversity (Szaro et al., 1998). Logging of old growth forests, conversion of land to agriculture, introduction of exotic species, and in particular, suburban sprawl have all contributed to the continued degradation of ecological systems. The fragmentation of habitat from outwardly expanding, low-density development across American landscapes is considered to be the leading cause of species decline and the loss of ecosystem integrity (Peck, 1998). From 1991 to 1998 alone, more than 80 percent of new housing construction took place in suburban communities (Hoffman, 1999). In almost every major metropolitan area across the Country, the amount of land consumed by development far outstrips the rate of population growth.

As sprawling development patterns continue to eat away at our landscapes, the critical natural resource base upon which we depend is showing signs of irreversible decline. A study estimated that 85 percent of endangered species in the US are threatened by habitat degradation (Dobson et al., 1997). Only 45 percent of the original naturally occurring wetlands remain in tact; 81 percent of fish communities nationwide are adversely affected by human development; and less than 2 percent of streams are of high enough quality to be worthy of federal designation as wild or scenic rivers (Noss and Cooperrider, 1994). Ecosystem degradation and the loss of biodiversity have been most pronounced in the South, Northeast, Midwest, and California where population growth and land development pressures have been the greatest (Noss and Scott, 1997).

A traditional species by species approach to regulation and management has been unable to solve many of the complex environmental problems facing the US (Yaffee and Wondolleck, 1997). Resource managers are quickly discovering that adequate levels of protection can not be achieved through a narrow set of policies and a fragmented development review process focused on individual parcels of land (Marsh and Lallas, 1995). Sustainable resource management must instead recognize the connections to broader ecological systems that extend beyond individual ownership, management jurisdictions, and even international boundaries (Christensen et al., 1996). Policies, for example, must acknowledge the fact that highway construction inland can destroy fish spawning areas along the coast; sewage outflow into a bay can negatively impact seabird breeding areas hundreds of miles away; and over-harvesting of fisheries in Canada ten years ago, may ruin this year's lobster harvest in New England.

A New Management Approach

In response to the increasing decline of critical natural resources across the US, public decision makers are abandoning the traditional species by species approach to regulation and instead are embracing ecosystem approaches to management. Ecosystem management represents a departure from traditional management approaches by addressing the interaction between biotic and abiotic components within a land or seascape, while at the same time incorporating human concerns (Grumbine, 1994). In this approach, entire ecological systems, and the ecological processes within them, become the framework for management efforts. Both academic researchers and public policy makers have proposed ecosystem management as a new "paradigm" of management and an improved framework for protecting resources over the long term (Cortner and Moote, 1999). At least 18 federal agencies have formally committed to the principles of ecosystem management and are exploring how this concept can be incorporated into their present day activities (Haeubner, 1998). The most recent comprehensive survey identified over 600 ecosystem management projects ranging from the Greater Yellowstone Ecosystem (GYE) and the Everglades Ecosystem to the Chesapeake Bay and the Gulf of Maine (GOM) (Yaffee et al., 1996).

Planners and resource managers increasingly recognize that while ecosystem management requires looking beyond specific jurisdictions and focusing on broad spatial scales, the approach will in part be implemented at the local level with local land use decisions. Furthermore, ecosystem approaches to management may not be realized solely by structural or engineering approaches to management, but by the coordination of local plans and policies across larger landscapes (Kirklin, 1995; Beatley, 2000). Local level planning therefore must be considered along with other spatial and jurisdictional scales when it comes to managing regional ecological systems. Many of the factors causing ecosystem decline, such as rapid suburban development and habitat fragmentation occur at the local level and are generated by local land use decisions (Noss and Scott, 1997). The vast majority of these decisions affecting large ecosystems will be made at a smaller scale where they make the largest impact on the natural environment (Endter-Wada, 1998; McGinnis et al., 1999). Local planning and decision-making for ecosystem management is perhaps most relevant in places where a fragmented pattern of land ownership is combined with strong development pressures.

As a consequence, some of the most effective policy tools that can either threaten or protect ecosystems are in the hands of county officials, city councils, town boards, local planning staff, and the general public rather than federal or regional agencies. Thoughtful policies and actions at the local level can often protect critical habitats of regional significance more effectively and less expensively than the best intentioned state or federal protection schemes (Duerksen et al., 1997). The importance of local ecosystem-based planning is further highlighted by the declining role of the federal government in the protection of habitat and associated ecological systems over the past 15 years, and a future political climate that suggests giving increasing control to local jurisdictions when it comes to making natural resource use decisions.

While much research has been geared toward instituting the broad principles of managing natural systems, comparatively little work has been done to evaluate the specific tools and strategies involved in implementing ecosystem management at the local level. To date, little or no systematic, empirical research has been done to inform local jurisdictions, such as towns, cities, and counties on how to best incorporate the principles of ecosystem management into their planning and regulatory frameworks. Ecosystem management was derived from federal-level thinking, but effective implementation of the approach will be achieved at the local level with sound planning efforts. Long-term success of ecosystem approaches to resource management thus rests on understanding how local plans effectively capture their key principles and practices.

The disconnect between federal-level thinking and local-level actions is the main problem addressed by this book. It was written in recognition of the shortcomings associated with ecosystem management in the US and the general lack of empirical research in addressing these problems. Its principal focus is on how local communities can effectively address ecosystem level problems through land use planning frameworks. While ecosystem approaches to management take place at a variety of geographical scales and jurisdictional levels, this work concentrates almost exclusively on the role of local jurisdictions comprised of a mosaic of private and public land ownership. Understanding the degree to which local communities incorporate the principles of ecosystem management into their plans and planning processes will provide guidance for planners, resource managers, and public officials on how they can protect ecosystems and their components into the future.

The examples provided and conclusions made throughout this book bring together five years of empirical research in Florida on the topic of how local jurisdictions can incorporate into their plans the principles underlying ecosystem approaches to management. The following chapters seek to answer specific questions, including: 1) What are the main components of or best practices for a sound ecosystem management plan at the local level; 2) Which state-mandated comprehensive plans are most geared to ecosystem management and why; 3) What are the factors and processes influencing the quality of comprehensive plans with regard to ecosystem-based management; 4) What motivates the implementation of plans and policies over the long term; and 5) How can plans, planning processes, and the state growth management programs that mandate them be improved to better accomplish ecosystem management? These questions are addressed using a variety of analytical techniques, including multivariate and spatial regression analysis, Geographic Information Systems (GIS), surveys, and case studies. In this way, the book examines the role local jurisdictions can play in managing broader ecological systems at a level of detail that has never been previously studied.

Research Setting: Why Florida

Florida has a combination of institutional, regulatory, and biogeographical characteristics that make it an ideal setting within which to examine ecosystem planning at the local level (Figure 1.1). A growing emphasis on ecosystem approaches to management alongside a strong land use planning regulatory framework create an atmosphere of high commitment and opportunity to implement the principles of ecosystem management. Add to the mix a national crown jewel of an ecosystem (i.e. the Everglades), fragile coastal environments, and intense pressure for urban and suburban development across much of the state and the result is an ideal situation to test the feasibility of using local planning to advance the concept of ecosystem management. Lessons learned from Florida's experience can be extended to other areas with similar characteristics across the nation and throughout the world.

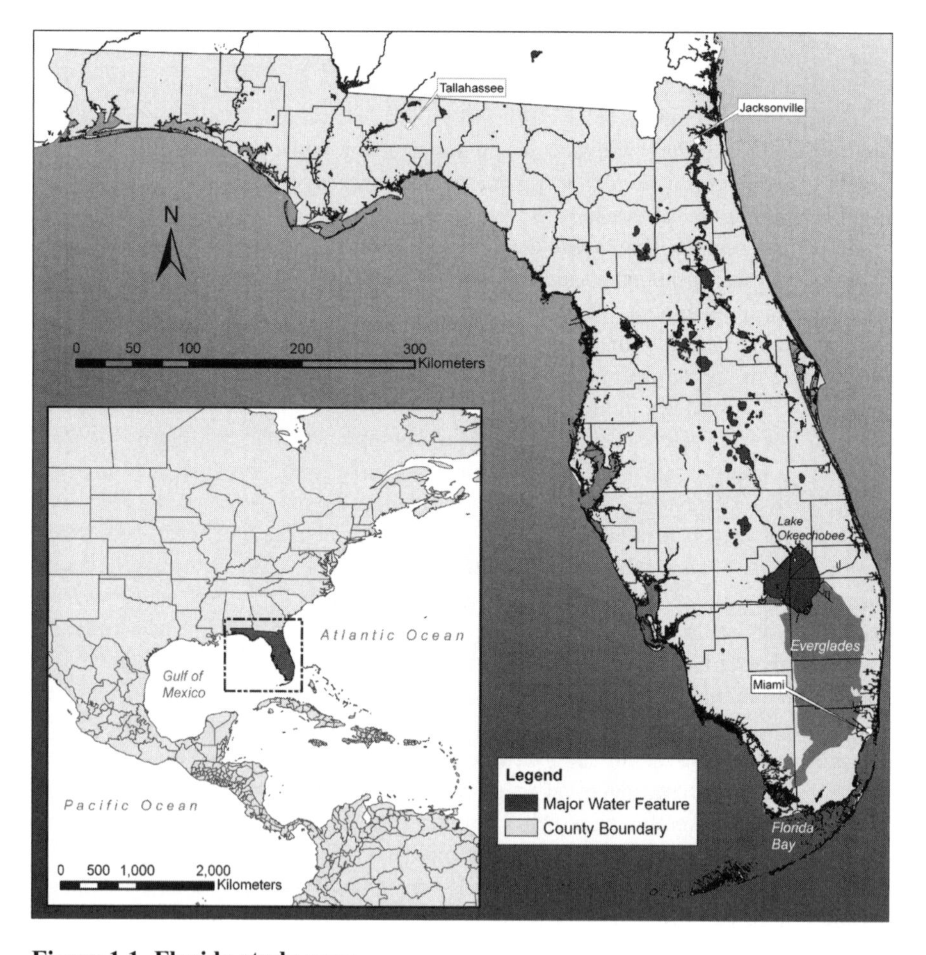

Figure 1.1 Florida study area

Source: Environmental Planning and Sustainability Research Unit, Texas A&M University.

Existing Biodiversity and Critical Habitats

Florida contains some of the most biologically diverse and valued ecosystems in the Nation and is widely recognized as one of North America's most important reservoirs of biological diversity (Cox et al., 1994). The State has more than 3,000 native plant species, distributed within 81 plant communities and approximately 668 species of native terrestrial and freshwater vertebrates (of which 17 percent are endemic). The centerpiece for Florida's biodiversity is the Everglades ecosystem. This 5,000 square-kilometer wetland system in southern Florida contains one of the highest concentrations of species vulnerable to extinction in the US. It is home to a great variety of key species including the American crocodile, Florida panther, and West Indian manatee. In addition to rare and endangered species, the Everglades are renowned for its abundance of birds, with 347 recorded species. The mangrove estuaries of Florida Bay, in particular, are a breeding habitat for Roseate Spoonbills, Wood Stork, White Ibis, Glossy Ibis, and eleven species of egrets and herons. Due to the prominence of the Everglades ecosystem, other critical natural resources of regional significance are often overlooked. For example, Sarasota and Tampa bays are considered estuaries of national significance; Kissimmee River north of Lake Okeechobee is still considered a key riparian area; and one of the last continuous swaths of biodiversity owned by private interests in the southeast US is situated in the Panhandle region of the State.

In recent decades, however, Florida has been noted not for its impressive diversity of species, habitats, and ecosystems, but rather the rate of their decline from human development. The Everglades area retains less than 10 percent of its original habitat as the human population density of southern Florida threatens to overrun one of the most unique habitats in North America. Only 5 percent of the original wading bird population remains and over 68 plant and animal species are threatened or endangered. Canals and levees capture and divert its water for human use, including drinking water, irrigation, and flood control. Often, too much water is withheld from the Everglades during the wet season, or too much is diverted into it during the winter drought, disrupting the natural cycles of feeding and nesting which depend on these patterns. Consequently, the region is experiencing unnatural water shortages and saltwater intrusion from over-pumping of groundwater supplies. Much of the time the water is contaminated by pollutants from upstream activities. As a result, over 1 million acres of the ecosystem are under health advisories for mercury contamination.

Decline of natural ecosystems is not limited to the Everglades. The growth of Florida's resident (over 600 new people added per day) and tourist populations, as well as its agricultural industry has contributed to a dramatic loss of forest and wetland communities across the entire state. Florida's native species are also suffering. According to the US Fish and Wildlife Service's rankings, there are 70 endangered (not including threatened, etc.) species in Florida, ranking it second in the nation, behind California. For every native species that is lost, ten non-native or invasive species have become established. Indicator species, such as the black bear have been heavily impacted by the increase in road construction (Cox et al., 1994; Beatley, 2000). Vacation home and resort development have had both direct and indirect impacts on species, such as the Key deer. Critical wetland habitats continue to be developed to make way for shopping malls, office parks, and resort communities. The precarious balance between rapid urban growth and the conservation of critical natural resources in Florida make it an ideal living laboratory within which to study the impacts of local land use decisions on protecting ecological systems.

State Focus on Ecosystem Management

Florida has a well-established framework for ecosystem management to ensure a level of consistency in the way the concept is understood and carried out. Local communities across the state seeking to protect broader ecosystems thus have a model on which to base their specific programs. In 1993, Florida's Department of Environmental Protection (DEP) recognized that traditional approaches to management were not able to adequately protect biodiversity and decided to reorient the state's environmental programs around an ecosystem approach to management. The DEP established the following four cornerstones to ecosystem management: 1) place-based management, 2) cultural change (in the agency's attitudes and practices), 3) commonsense regulation, and 4) improved foundations of science and technology on which to base decisions (DEP, 1995a). Under this approach, DEP has moved away from media-based management, which addresses water, air, and land separately, and toward an integrated understanding of problems and solutions based on natural boundaries rather than those defined by humans. Furthermore, DEP is attempting to redefine public and private roles in managing resources by encouraging public involvement, seeking public consensus, and striving to develop an ethic of shared responsibility for the natural environment. Ecosystem Management Areas (EMAs) have been defined as the natural units within which programs, activities, and specific plans are to be directed (DEP, 1995b).

In the late 1990s, DEP publicly abandoned its EMA framework but continued to pursue its ecosystem management program through a watershed approach to planning. The Watershed Management Program was created in October of 1999 to implement the provisions of the Florida Watershed Restoration Act of 1999 (Section 403.067, Florida Statutes). DEP coordinates planning for watershed units (essentially EMAs) with local government and stakeholder groups. The EMA concept still persists among local groups, however. For example, a multi-jurisdictional and stakeholder management plan based on the Lake Worth Lagoon EMA in Palm Beach County was revised in 2007 and continues to drive habitat restoration projects within that designated ecological unit.

Perhaps the most renowned ecosystem effort is the Comprehensive Everglades Restoration Plan (CERP), a 30 year, $10.9 billion (originally funded at $7.8 billion) project to restore the water flows associated with the Everglades ecosystem to a more "natural" state. The Plan is primarily focused on structural engineering approaches to re-pipe the everglades system, which originally consisted of a slow moving sheet flow of water from Lake Okeechobee south into Florida Bay over an 18,000 square-mile area. Projects include surface water storage reservoirs, aquifer storage wells, treatment facilities, new canals, etc. The Plan was approved in the Water Resources Development Act (WRDA) of 2000 (CERP) and is currently being funded, managed and implemented through a 50/50 partnership between the state and federal governments. While the restoration plan is often criticized for leaving out the people and the policies necessary for effective ecosystem approaches to management, it has at least served to peak the interest of the 7 million people (which is expected to double in the next 50 years) covering 16 counties that will be potentially impacted by the project.

These are just a few examples of the myriad of ecosystem-based projects in Florida taking place at multiple levels of government and non-government organizations. None of these projects focus on local jurisdictional land use planning strategies to balance rapid population growth with protecting critical natural habitats. However, they certainly serve as external influences on local-level decision-making and provide a greater opportunity for implementing the principles of ecosystem management at the local level throughout Florida.

The Local Comprehensive Planning Framework

Florida requires that each local community (cities and counties) prepare a legally binding comprehensive plan. While there are many different types of resource management plans in Florida (some of which are described above), comprehensive plans follow a consistent format (in terms of production, element types, and review/updating processes), are an institutionalized policy instrument dating back several decades, and most importantly provide a basis for city and countywide land use and resource management decisions. In this sense, comprehensive plans are an important tool for accomplishing many of the goals of ecosystem management because they are the starting point for specific ordinances, land development codes, and environmental policies. They also often incorporate and implement more regional environmental activities, such as National Estuary Programs (NEP), EMA plans, and other agreements on transboundary[1] resource management. Most importantly, comprehensive plans in Florida are where the "rubber hits the road" when it comes to managing critical natural habitats and ecological processes over the long term. If ecosystem approaches to management are going to be effectively implemented, they must be rooted in the local policies guiding development decisions.

City and county comprehensive plans in Florida stem from the 1985 *Local Government Comprehensive Planning and Land Development Act*, which mandated that new local comprehensive plans be written and required that they be consistent with goals of the State land use plan. The broad mandates of the growth management legislation (meant to upgrade 1975 legislation) were given shape and substance by Rule 9J-5, which sets minimum standards for judging the adequacy of local plans submitted to the state for approval. Rule 9J-5, adopted by the Department of Community Affairs (DCA) in 1986, requires that specific elements be included in local plans and prescribes methods local governments must use in preparing and submitting plans. Required elements, among others, include land use, coastal

1 The term transboundary is defined for this study as a management approach that focuses beyond a single human boundary, such as a local jurisdiction or some line of human ownership.

management (where applicable), conservation, and intergovernmental coordination. In each element, the rule lists the types of data, issues, goals, and objectives that must be addressed using a "checklist" format (May et al., 1996). For example, in the conservation element, objectives must conserve wildlife habitat while policies must pursue cooperation with adjacent local governments to protect vegetative communities (9J-5.013). Many of the required goals, objectives, and policies contained within a comprehensive plan lay the foundation for ecosystem management at the local level. Rule 9J-5 also sets forth requirements on public participation throughout the planning process (9J-5.004).

At the heart of this coercive and highly detailed state-planning mandate lies the requirement for each local jurisdiction to adopt a future land use (FLU) map. This "regulatory and prescriptive" map designates the types of land uses permitted in specific areas within each local jurisdiction. The requirement is meant to ensure that growth and development proceeds with adequate public infrastructure, does not adversely impact critical natural habitats (e.g. wetlands), and does not promote the harmful effects of urban and suburban sprawl. Each adopted plan under the State mandate is thus a legally binding policy instrument offering spatial guidance for future development patterns. It is not just a broad, strategic policy statement, but a set of explicit directives adopted through a participatory planning process where future outcomes are expected to conform to the original design of the plan.

The 1985 Act was updated in 1993, but still remains as the primary instrument driving local resource and land-use decisions. While the prescriptive and coercive nature of the planning legislation has been questioned in recent years (May et al., 1996; Burby et al., 1997; Catlin, 1997), it provides an ideal setting in which to assess the effectiveness of local plans in achieving the principles of ecosystem management. The required elements, objectives, and policies create a standardized, comparable framework for implementing ecosystem management at the local level. Although the state mandate does not explicitly require ecosystem approaches to management in local comprehensive plans, in most cases written elements call for the management of ecological systems that extend beyond a local jurisdiction. For these reasons, comprehensive plans in Florida are the principal unit of analysis for this book and serve as the basis for examining how local communities can accomplish the principles of ecosystem management.

Organization of the Book

This book is organized into four parts containing 13 chapters. Succeeding the Introduction (Chapter 1), Part 1 focuses on the local plan as an instrument for managing the natural environment, specifically ecological systems such as watersheds. Chapter 2 reviews several literatures to form a better understanding of the major principles of ecosystem management and to lay the foundation for a conceptual framework for local ecosystem plan quality. The following three major areas of literature and associated concepts are examined: ecological science, organizational design, and collaborative planning. Chapter 3 builds on these concepts to derive a set of integrated principles that encapsulate the major themes of ecosystem management.

These principles synthesize the various literatures and help guide the reader toward understanding what makes for effective ecosystem approaches to management from a local planning perspective. Chapter 4 examines the literature on plan quality as a basis for conceptualizing and measuring ecosystem management capabilities at the local level. It then ties together the principles of ecosystem management plan quality through the development of a plan coding protocol used to measure local ecosystem plan quality. This protocol captures the essential elements of ecosystem management through the core components of conceptualizing plan quality. Chapter 5 presents the results of two case studies in Florida. The first evaluates the plan coding protocol against a random sample of local jurisdictions. The second uses GIS and spatial analysis to map and measure the mosaic of local environmental policies across large landscapes.

Part 2 of the book focuses on factors influencing local ecosystem plan quality. Chapters 6 and 7 examine the role of the biophysical landscape and its components, particularly biological diversity (biodiversity), in shaping the environmental content of local plans. The results from two case studies are presented. The first measures the amount of biodiversity within a local jurisdiction and explains its impact on the quality of local plans. The second study employs GIS mapping techniques to analyze adjacent jurisdictions in the southern part of Florida as a way to better understand how the level of biodiversity and human disturbance on critical natural resources motivate planners to adopt high quality plans. Chapters 8 and 9 look at the impact of public participation and stakeholder representation on local ecosystem plan quality. An in-depth case study is presented examining the influence of stakeholder participation in ecosystem approaches to management.

Part 3 of the book concentrates on the implementation of local plans and policies associated with ecosystem management in Florida. Chapter 10 examines the importance of implementation in environmental planning and the lack of attention it receives from practitioners and in the planning literature. Chapter 11 highlights two case studies that use novel approaches to map, measure, and predict the degree local plan implementation within large watersheds.

Part 4 discusses local planning implications from the results in the preceding chapters and presents a set of recommendations to improve the process and practice of environmental planning at the local level. Chapter 12 lists an integrated set of recommendations stemming from empirical results covering the planning process, the art of plan making, and the implementation of adopted plans. Chapter 13 summarizes key findings on how local communities can more effectively plan for and manage surrounding natural systems over the long term.

PART 1
The Plan: Conceptualizing and Measuring Local Ecosystem Plan Quality

Part 1 of the book focuses on the local plan as an instrument for managing human activities impacting the natural environment, specifically ecological systems such as watersheds. As shown in Figure 2.1, this part of the book begins by reviewing several literatures and concepts underlying the practice of ecosystem management (Chapter 2). Based on this analysis, a set of integrated principles capturing the major themes of ecosystem management are presented (Chapter 3). Chapter 4 examines the literature on environmental plans and plan quality as a policymaking vehicle for conceptualizing and measuring ecosystem management capabilities at the local level. The principles of ecosystem plan quality are synthesized through the development of a plan coding protocol used to measure local ecosystem plan quality in subsequent analyses (Chapter 5). This protocol captures the essential elements of ecosystem management through the core components of plan quality. Finally, Chapter 5 tests the effectiveness of this plan coding protocol by evaluating it against two samples of local comprehensive plans throughout Florida. Readers who wish to skip material on the theoretical and conceptual underpinnings of local ecosystem plan quality can move directly to Chapter 5.

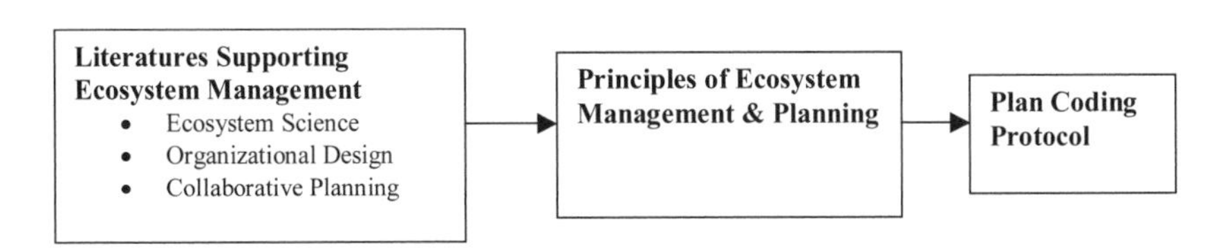

Figure 2.1 Developing a conceptual definition for ecosystem plan quality

Chapter 2

Analysis of the Literatures and Concepts Underlying Ecosystem Management and Planning

Because ecosystem management requires a holistic, interdisciplinary, and boundary spanning approach to decision-making, it is necessary to draw upon a variety of literature bases to define its key conceptual components. This chapter reviews several literatures to form a better understanding of the major principles of ecosystem management and to lay the foundation for the ecosystem plan quality protocol described in Chapter 4. After a brief overview of the development of ecosystem management as a policy-making paradigm, the following three major conceptual areas are examined: ecological science, organizational design, and collaborative planning. These literatures provide a theoretical social and natural science foundation that support the practice of ecosystem management. By examining the contributions of each literature associated with ecosystem management, the defining core components of this resource management approach can be more easily distilled.

Overview of Ecosystem Management: The Development of a Natural Resource Management Paradigm

Ecosystem management is often treated as a modern-day concept and a next generation approach to environmental policy (Estay and Chertow, 1997). However, the idea was first developed over 50 years ago by Aldo Leopold who wrote that people should take care of the land as a "whole organism" and try to keep all of the cogs and wheels in good working order (Leopold, 1949). Leopold recognized many of the interdisciplinary principles of ecology, economics, and other human interests associated with managing natural systems today. Even before Leopold, the less well-known Ecological Society of America's Committee proposed the concept for the Study of Plant and Animal Communities, which recommended protecting ecosystems as well as individual species (Shelford, 1933). In the late 1970s, the grizzly bear and northern spotted owl controversies coalesced the different historical aspects of ecosystem management into an applied holistic practice (Szaro et al., 1998). The failure to adequately protect these key species caused resource managers to look more broadly at entire ecological systems and their interrelated habitats as a more effective strategy for maintaining species populations. Intense media attention and controversy over the decline of popular species further expedited the movement to embrace the concepts of ecosystem management.

Today, ecosystem management is a broadly accepted and commonly used term in both academic and governmental arenas. While this approach to management has become a major focus for future environmental policy in the US, individuals and organizations interpret it differently. A leading literature review by Grumbine (1994) defined ecosystem management as "integrating scientific knowledge of ecological relationships within a complex sociopolitical and values framework toward the general goal of protecting native ecosystem integrity over the long term." In contrast, the American Forest and Paper Association (1993) defines the term as "a resource management system designed to maintain or enhance ecosystem health and productivity while producing essential commodities and other values to meet human needs and desires within the limits of socially, biologically, and economically acceptable risk." In fact, there are a vast array of definitions proposed for ecosystem management (Christensen et al., 1996; Czech and Krausman, 1997) indicating the amorphous nature of the concept and its practice in the US and worldwide. While each definition has a slightly different nuance or focus, most contain the following central elements: 1) protection of ecological integrity, functions, and process; 2) intergenerational sustainability; and 3) incorporation of human needs, values, and interests.

Regardless of the specific definition, it is widely accepted that an ecosystem approach to management is distinguished because it takes a holistic approach to addressing natural resource issues by focusing on the interaction between human communities and entire ecological systems (Grumbine, 1994). It attempts to transcend jurisdictional lines by broadening managers' geographic focus and by creating situations of collaborative problem solving. This management framework is based on an ecosystem science that integrates many disciplinary approaches and addresses ecological issues at sometimes very large temporal and spatial scales (Szaro et al., 1998). Ecosystem management is thus a place-based concept that focuses on the boundaries of ecological systems rather than on traditional jurisdictional lines. By aligning policies and plans with a coherent spatial unit, ecosystem approaches to management can more effectively protect ecological structure, function, and overall biodiversity. There are many different ways to conceptualize an ecosystem, making it a vague and elusive concept for resource managers. A dung pile is as much an ecosystem as a watershed, or even the entire eastern seaboard of the United States. The scale at which an ecosystem is drawn depends on how environmental problems are perceived and is in many ways a political issue. To further complicate matters, ecosystems are dynamic, constantly changing, and vary continuously along gradients of space and time (Allen and Hoekstra, 1992; Haeubner, 1998; Brussard et al., 1998).

By focusing on ecological boundaries as opposed to those defined by humans, ecosystem management differs in many ways from traditional resource management policies and practices (Table 2.1). Traditional resource management tends to view nature in terms of maximizing consumption for human uses. Decisions are primarily in the hands of scientific experts who consider management more of an engineering problem than a balancing act between human interests and maintaining the functions of ecological components. This approach to management often gives rise to top-down models of governance that promote rational or instrumental forms of planning.

Community participation and the incorporation of values in the decision-making process are not emphasized.

In contrast, ecosystem approaches to management usually view nature as an open system that is constantly changing over time. Ecological relationships are complex and often unknown so that management must be adaptive to sudden shifts in equilibrium. Humans and their values are considered a part of the nature system and are directly incorporated into the decision-making process. Ecosystem management promotes bottom-up models of governance, where collaboration and discourse among multiple stakeholders is essential to an enduring policy outcome. Most notably, it makes ecological sustainability – long-term maintenance of ecosystem productivity and resilience – a primary goal of resource management.

Ecosystem management also recognizes a critical interdependence between social and ecological vitality, including humans and human societies in resource management to an unprecedented extent. It breaks new ground by making the social and political basis of natural resource management goals explicit and by encouraging their development through an inclusive and collaborative decision-making process (Cortner and Moote, 1999). Underlying this management approach is the assumption that when faced with the threat of environmental degradation, individuals will not always follow rational choice models of decision-making, but instead will work collectively to solve transboundary resource problems (Ostrom, 1990). The practice of ecosystem management is thus transboundary, interagency, multi-party, and interdisciplinary (Yaffee, 1996). This approach to decision-making is so difficult because it hinges upon a range of thorny issues technical planners often wish to ignore, such ecological understanding, organizational structure and design, intergovernmental collaboration and planning, private ownership, and even individual values.

As illustrated in Table 2.1, ecosystem management is an interdisciplinary approach to decision- making that draws upon several areas of literature. To thoroughly understand its principal themes and how they can be captured in a local land use plan, we must look beyond the broad-based work on ecosystem management and focus more closely on the major literatures supporting this brand of decision-making. The following sections investigate what I identify as the three major literatures underlying ecosystem approaches to management: ecosystem science, organizational development, and collaborative planning. For each literature, a brief background on its traditions is provided followed by its major contributions for determining principles of ecosystem management and a conceptual definition for ecosystem plan quality.

Table 2.1 Comparison of key concepts in traditional resource management and ecosystem approaches to management

	Traditional management	Ecosystem management
Goal	Manipulate natural resources to optimize productivity for human use	Maintain the integrity of ecological systems through an ecologically sustainable approach to management
Nature	A collection of resources to be consumed	Complex, constantly changing, interrelated systems to be used sustainably for future generations
Ethics	Compartmentalized; interrelationships marginal	Holistic; interrelationships important
Science and Models	Deterministic, linear, static; full knowledge required, approaching steady-state equilibrium	Stochastic, nonlinear, dynamic; variable-rate dynamics with temporary equilibria upset periodically by chaotic moments that set the stage for the next temporary equilibrium
	Robust, well-defined theory; discrete data with predictable outcomes	Beginnings of theory; theory and practice intertwined, interrelated data with unreliable outcomes; surprises to be expected
	Maps, linear optimization, monetized cost-benefit analysis, natural science-based	Geographic Information Systems (GIS), related databases, nonlinear simulation, evaluation for social, economic, political aspects
Management and Organization	Centralized; rigid; little acceptance of incentives	Decentralized; interrelated teams; adaptive and flexible; focus on incentives, innovation, and collaborative learning
	Hierarchical, top-down bureaucracies	Adaptive, bottom-up, open, collaboration-based
Planning	Comprehensive, rational	Interrelated, communicative, ends not yet fixed
Decision-making	Rigid, authoritarian, reliance on experts/technicians	Discourse-based, inclusion of all key stakeholders
	Science driven	Science provides information, but does not drive the decision- making process; externalities considered
Participation	Little participation	Deliberative, stakeholder participation essential
Leadership	Authoritarian; leaders chosen	Situational; leaders emerge from the community of stakeholders when needed

Source: Adapted from Cortner and Moote (1999) and McCormick (1999).

Ecosystem Science and Management

While the literature on ecosystem science and management is quite diffuse, its main contribution to an ecosystem approach to planning comes from its emphasis on natural science and ecological understanding. Numerous publications on resource management and the ecosystem concept emerged in the late 1980s and extend through the 1990s. Specifically, the literature focuses on ecological concepts essential to understanding ecosystem function and the structure of natural system. Research and writing from this area provide an ecological justification for the various policies, plans and projects that seek to manage ecological systems over the long term.

The literature on ecological science and ecosystem management identifies three key principles managers must consider when constructing social science frameworks to address ecosystem-related problems. These principles are summarized by a key article (Christensen et al., 1996) to the Ecological Society of America, which has helped create consensus among scientists on how ecosystems function. This work helped set the stage for linking broad management frameworks to the underpinnings of ecosystem science. The following key principles help inform a definition of ecosystem management plan quality:

1. Broad spatial and temporal scales: Ecosystem function includes inputs, outputs, cycling of materials and energy, and the interaction of organisms that all operate at different spatial scales (Cortner et al., 1998). Boundaries defined for the study or management of one process are often inappropriate for the study of others. For example, a management issue, such as developing a recovery plan for an endangered species may present itself at one scale of organization, but a complete understanding and resolution of the issue usually requires integration across several scales and levels of organization. When developing ecosystem policies, it is therefore important to acknowledge processes operating at lower and upper levels of organization. The non-hierarchical nature of organizational levels means that ecosystem management must take a population, community, and ecosystem perspective simultaneously (Brussard et al., 1998; Allen and Hoekstra, 1992). A complicating factor when it comes to actually implementing this principle is the fact that local jurisdictions rarely follow natural boundaries but rather some arbitrary administrative unit. This drawback is a source of constant frustration for ecosystem planners. Another factor complicating management initiatives is the fact that ecosystems operate over wide temporal scales. Short-term policies that focus on several years are ineffective in addressing ecosystem processes that occur over decades and centuries (Lessard, 1998; Cortner et al., 1998).
2. Structure, diversity, and integrity: Ecosystem function depends on its structure, diversity, and integrity. Management must recognize that biological diversity and structural complexity strengthen ecosystems against disturbance and supply the genetic resources necessary to adapt to long-term change. An ecosystem that is undisturbed by human actions will have a high level of integrity and be able to maintain its structure, species composition, and disturbance regime in a self-sustaining fashion (Brussard et al., 1998; Peck, 1998). However, since

almost all ecosystems are disturbed at some level by human activities, the goal for policy makers is to maintain the remaining critical functions and processes that support the overall system (for an additional discussion see Chapter 7). This approach is often translated into protecting the mosaic of critical habitats that support metapopulations over large areas.

3. Ecosystems are dynamic: Ecosystems are constantly changing over space and time. Both successional processes and human disturbance contribute to the dynamic nature of ecosystems so they never reach a point of equilibrium or stability (Lackey, 1998; Vogt et al., 1997). Ecosystem change can be measured on a timeline of hundreds if not thousands of years.

Based on these three principles of ecosystem science, several conclusions can be made for understanding what makes for effective ecosystem management. Ecosystem management should consider entire ecological systems at various levels of organization, rather than focusing on a single fragment. While looking at the big picture, the approach should concentrate on protecting critical habitats that support the function, structure, and integrity of the natural system. Finally, the management approach should be adaptive to respond to constantly changing ecological conditions. Implementing these concepts into a local plan translates into the following. First, the plan must be able to look at large areas that coincide with an ecological unit. If this unit extends beyond a single jurisdiction (thus becoming transboundary), the plan must have the capability to coordinate with other jurisdictions and/or organizations. Second, the plan must be long-range. It must set a vision for the future and put forth broad goals that it expects to achieve over time. Some local plans in place already project 25 to 50 years into the future. Third, the plan must be focused in some way on identifying and managing or conserving critical ecosystem components over the long term. Finally, the plan must be adaptable. That is, it must be able to be updated continuously as ecological conditions change over time.

While the literature on ecosystem science and management provides a foundation for understanding how natural systems work, it cannot solely be relied upon to derive a global set of principles of ecosystem management or identify what makes an effective ecosystem plan. The major shortcoming of this literature is that it follows a strict natural science or technical rational approach to management that tends to leave humans, human values, and human planning processes out of the planning and management picture.

Principally, the ecosystem science and management literature assumes a high level of ecological understanding. Many scientists assert that effective ecosystem management must be based on a comprehensive understanding of ecological principles (Reichman and Pullman, 1996). The view that successful management should rest on improving our knowledge of ecological processes leads to several problems. Scientific understanding will never be complete or even to the point that management decisions can be made with scientific certainty (Sexton, 1998; Holling, 1996). Also, lack of ecological knowledge is often used as an excuse to prolong expensive studies in lieu of taking protective actions. It may be better to act with incomplete information than not act at all when complete information is unattainable.

Overemphasizing the role of science and ecological understanding creates a planning process that is led by scientific facts rather than human choices. The literature suggests that processes are based on a linear, technically rational approach to management where the focus is on collecting scientific data, fixing goals, and making decisions through expert advice. Ecosystem planning, however, is "not a technical problem" that can be solved through the manipulation of nature (Gerlach and Bengston, 1994; Brunner and Clark, 1997). Rather, it involves making collective choices based on a set of values that affect humans and their relationships with the natural environment over the long term (Stanley, 1995; Endter-Wada, 1998). As such, humans and their interests must be considered integral components of an ecosystem, rather than disturbances in an otherwise smoothly running system (Williams and Stewart, 1998).

In this sense, ecosystem management should be a social process that is cognizant of and interested in sound ecological science, but which is driven by a search for deeply held, culturally rich connections between local communities and their place (Norton, 1998). Perhaps Salwasser (1994) said it best in his constantly borrowed statement that ecosystem management "is more about people than anything else." For insight on the human aspects of maintaining the integrity of natural systems, it is thus necessary to review other literature bases from the social sciences.

Organizational Design

The organizational design literature examines collaboration from an organizational standpoint. It seeks to better understand how one or several organizations, such as state or local governments can facilitate collaboration around multiparty problems. Organizational design's major contribution to developing principles for effective ecosystem management is that scholars provide key insights into how to promote collaboration among institutions and jurisdictions to accommodate the management of complex natural systems. Many of the core themes coming out of organizational theory, particularly from a business perspective, such as stakeholder networks, systems thinking, and learning organizations help form an understanding of effective ecosystem approaches to management that can be then captured in a land use plan. The following sub-sections elaborate on the core organizational concepts supporting ecosystem management.

Inter-organizational Collaboration

Ecosystem approaches to management increasingly depend on collaboration across political, administrative, and ownership boundaries (Blumenthal and Jannink 2000; Selin et al., 2000). Because ecosystem approaches to management adhere to ecological systems (and associated communities of interest), rather than administrative or political lines, inter-organizational collaboration across jurisdictions, agencies, and land ownership is necessary to achieve effective management of transboundary resources. Ecosystem management from this perspective thus seeks to manage how humans impact resources and compel human institutions to coordinate their efforts toward common goals. The organizational design literature provides a solid theoretical starting point for understanding how collaboration can lead to more effective management of transboundary ecological systems.

Scholars in this area have, over the years, increasingly called for inter-organizational and intersectoral collaboration to solve major environmental problems (Gray, 1989; Westley, 1995; Cortner et al., 1998; Cortner and Moote, 1999). Such collaborations, which include a variety of public-private partnerships, alliances, and networks, have been viewed as critical to effective management outcomes that meet the needs of all interested parties (Gray and Wood, 1991; Westley and Vrendenburg, 1997). In this sense, collaboration induced by shared visions are intended to advance the collective good of the stakeholders involved (Bryson and Crosby, 1992).

Emery and Triste (1965) were one of the first in this tradition to argue that problem domains (ill-defined problems that depend on multiple perspectives for their solution) could be stabilized by inter-organizational collaboration. These so called "meta-problems" which transcend boundaries of single organizations (such as the management of ecological systems) must be addressed cooperatively. Perhaps the most influential work coming out of the business theory literature was Freeman's (1984) book on the stakeholder approach. This was the first major piece within an organizational context that suggested that not one, but multiple stakeholders are needed to solve complex problems. Freeman's thesis was that an organization or corporation must consider the interests of NGOs, government, and community groups in its planning. It called for a problem-solving approach that includes the input of many groups. This book, as a seminal work, touched off a barrage of articles and research focusing on organizational collaboration.

For example, many theorists promote the concept of organizational networks to achieve collaboration. Network structures occur when organizations realize that working independently is not enough to solve a particular problem or issue area (Khator, 1999). Typically, network structures form when people or organizations realize they are only one small piece of a total picture. It is the recognition that only by coming together to actively work on accomplishing a common mission will something be accomplished (Gray, 1989; Lee, 1992). The similarities between the way organizational theorists describe networks and ecologists explain the function of ecological systems is just beginning to be recognized (Wheatley, 1992). Networks of interconnected habitats (described in following sections) maintain ecosystems in a manner similar to the way organizations can solve more global ecological problems through a network structure. An organizational network that effectively mimics an ecological network or infrastructure may be a viable solution to managing ecosystems over the long term.

Jennings and Zandbergen (1995) take the next step by examining the role of organizations with respect to ecological sustainability. They offer an alternative to the traditional view that focuses on the single organization by proposing that a system or network of multiple organizations is the only way to facilitate sustainability over the long term. This is because managing natural systems requires addressing problems and places that extend beyond the domain of a single institution. Thus, "individual organizations do not contribute to sustainability as much as regional networks or organizations or industries that target ecosystems" (p. 1023).

Another important concept is "strategic bridging" where organizations act to bridge others so that pooling of resources and "interpenetration" of organizations occurs (Westley and Vrendenburg, 1997). Chisholm (1989) argues that where formal organizational arrangements are absent, insufficient, or inappropriate for providing requisite coordination, informal adaptations develop to satisfy that need. In this sense, informal methods compensate for the failure of formal structures to promote coordination. Informal organizations do not need to replace formal structures, but instead act as their complement, filling in gaps, strengthening ties, and providing a flexible, adaptive approach to collaboration.

While the literature shows collaboration can be beneficial when managing transboundary ecological systems, there are also arguments against collaborative arrangements (Coglianese, 1999; Conley and Moot, 2003). Bringing together multiple parties to solve common resource problems can increase conflict and reduce the chances a plan of action will be adopted. Even if solutions are agreed upon, the outcome of collaboration may be a watered-down or inappropriate management plan. In addition, collaboration around natural resource management can be expensive, time-consuming, result in a loss of control by government officials, reinforce negative stereotypes, and result in outcomes that meet only a few interests (Kennedy, 2000).

In Florida, it is largely recognized at the state level that ecosystem approaches to management are an important aspect of effective environmental management. Ecological systems, particularly regional watersheds, extend across multiple jurisdictions making sustainable management of the entire natural system more complicated (Kirklin, 1995). Because ecosystems often do not adhere to what has become a "crazy quilt" of land ownership and governance, environmental management goals are not being reached and natural systems such as the Everglades continue to decline (Light et al., 1995; Daniels et al., 1996). While this natural system is intricately connected over broad spatial and temporal scales, the land use decision-making framework is limited to local jurisdictions and some limited input from regional planning councils. Uncoordinated local land use decisions have cumulative negative impacts on the system as a whole. In other words, the Everglades and its sub-ecological systems are suffering a "death" from thousands of locally imperceptible, individual development decisions. Collaboration across jurisdictional lines and among multiple organizations thus becomes imperative if approaches to ecosystem management are to be attained (Daniels and Walker 1996; Randolph and Bauer 1999).

Systems Thinking and Learning Organizations

An additional concept leading to the definition and measurement of ecosystem management plan quality is called systems thinking. This concept comes from business-related organizational theory (this was originally a borrowed theory from systems engineering derived at MIT in the very early 1980s). Under this mode of thought, individuals working within organizations must acquire a systems thinking approach to management, where participants are able to step back and view the "big picture." According to Senge (1990) Systems thinking is a discipline for seeing wholes. It provides a framework for understanding the interrelationship, patterns, and structures (both human and ecological) underlying natural resource problems. Senge (1990) asserts that systems thinking requires "metonia," a shift of mind that enables the individual to see the larger processes at work and to see themselves not as helpless reactors, but active participants shaping their reality and creating the future.

There are clear applications of systems thinking to understanding and managing natural systems comprised of interrelated habitats that often extend across multiple jurisdictions, organizations, and landowners. Having the ability to look at the entire ecological system, even if it extends beyond a planner's jurisdiction is a critical aspect to effectively managing ecosystems. Systems thinking is thus a conceptual vehicle for sound ecosystem planning.

The manifestation of a systems approach to management is a learning organization (Westly, 1995). Management structures and their policies must be flexible and responsive to the constantly changing conditions of ecological and political systems (Lee, 1992). The learning organization, or in our case a planning agency, relies on adaptive management techniques, where policies are experiments to be tested and organizations are reflective units that improve their management capabilities over time (Daniels and Walker, 1996; Holling, 1996). To achieve these qualities, organizations rely on a process called double-loop learning, which requires rethinking the purposes and rules of operation to diagnose the problems of theory that underlie practical problems. In other words, members of organizations return to and challenge their core values by asking what they are attempting to accomplish in the first place. With double-loop learning, individuals look at the big picture inherent in systems thinking. This approach stands in contrast to single-loop learning, which is based on response to failure and adjustment of management strategies.

Organizational learning requires people to change some of their assumptions and views of how they understand the natural and institutional world. Working and thinking as a team, whether it is within a single organization or among several, thus becomes a central component of the learning process (Daniels and Walker, 1996). When it comes to managing ecosystems, which constantly change over time, a planning agency must develop the characteristics of a learning organization. The goal in this case is to prepare a plan that adapts to changing environmental and social landscapes. An effective ecosystem plan is therefore one that can be adjusted or updated to accommodate new scientific data or swings in community sentiment.

Organizational theory and design provides a basis for understanding inter-organizational collaboration, offers the beginnings of alternative institutional design models to deal with complex problems, and introduces the concept of social learning (embedded in planning theory) as a constant iteration of management, rather than a linear process. These are all important concepts that lead to a more thorough understanding of what makes effective ecosystem management and what makes a high quality ecosystem plan. To summarize, collaboration among jurisdictions, organizations, and landowners is essential to managing natural systems. The ability to take a big picture approach and look at the entire political and natural systems is also a key component in planning for critical natural resources. Finally, the capability to monitor and adapt to changing conditions by shifting the focus of a plan is a vital element in achieving effective approaches to ecosystem management.

However, this literature falls short when it comes to understanding how to accomplish sweeping changes and set up alternative organizational designs. First, while almost every expert is proposing organizational change to facilitate ecosystem management, few have provided clues on how to begin reform, where it should start in the organization, and how it should be carried out. Much of the debate is purely structural, such as whether change should involve a comprehensive overhaul or an incremental alteration of existing mechanisms. Few, if any, organizational theorists have looked beyond structure to the people working within agencies and institutions. This viewpoint reveals the difficulty in creating change. Removing ingrained standard operating procedures (SOPs), operational norms and long held values, and negative stereotypes is an extremely difficult task that receives little recognition in the literature. Government bureaucracies are notorious for being resistant to new approaches; those involved in resource management should be no exception.

Second, many authors call for collaboration, but provide little or no insight into how it can be successfully accomplished across fragmented ownership (Cortner and Moote, 1994). For example, Ostrom (1990) identifies a framework for collaboration in managing common pool resources, but focuses almost exclusively on the local level and within a single jurisdiction. Chisholm (1989) examines how collaborative networks operate across different organizations, but restricts his discussion to an existing organizational system already linked by function, product, and funding. Ecosystem management will often require coordination across various jurisdictions and between organizations that have no prior experience working together.

Third and most importantly, however, is that the organizational literature often leaves out the people and the process. While the organizational level is important, individuals ultimately drive it such that ecosystem planning is a product of individuals working together. To better understand how to bring people together and design processes with specific tools to facilitate collaboration and effective ecosystem management, we need to examine the field of collaborative planning. Planning, among other functions, looks beyond the façade of the organization to the people within them driving the quality of plans, policies, and solutions to natural resource problems. In this respect, collaborative planning provides a final conceptual link to developing an interdisciplinary, measurable definition of ecosystem management plan quality.

Collaborative Planning

The collaborative planning literature's primary strength in contributing to principles of effective ecosystem management and an understanding of what make a sound ecosystem plan is that it focuses on the people and the process of making sound decisions. As with the organizational design literature, the underlying notion is that management decisions must be made collectively because in most cases no single entity has jurisdiction over all aspects of an ecosystem. The need to integrate the values and knowledge of a broad array of organizations and individuals translates into a need to focus on collaborative planning efforts among resource owners, managers, and users (Cortner and Moote, 1999; Wondolleck and Yaffee, 2000). By drawing from the principles of conflict management and consensus building, the literature takes the first steps toward understanding how to bring a variety of interests together to solve complex, multiparty environmental problems. It provides tools and techniques to not only facilitate collaboration, but ensure that the final decision is enduring and implemented. In this sense, the collaborative planning literature instills values and personal preferences into the collaborative equation, providing key insights into how to accomplish effective ecosystem planning at the local level.

From a theoretical perspective, collaborative planning is rooted in the notion of communicative rationality. The goal of communicative rationality is to organize dialogue to promote democracy and personal growth, and search for a solution agreed upon in undistorted communication. The concept is based on the idea of achieving non-coerced mutual understanding and consensus among a community of inquirers through dialogue, where all have an equal opportunity to participate and make and challenge statements made by participants about what the community "ought to be." Consensus is thus reached through the force of a better argument.

Communicative rationality draws upon Habermas' (1979) critical theory and concept of communicative action. The idea behind communicative rationality as it is applied to collaborative planning is that it provides a forum for the local community to mutually debate, rationally consider, and reach consensus on public issues relevant to plan making. As Sager (1994) notes, a community can rationally achieve the goals to be collectively pursued. Values and norms, which could not be seen to have any rational founding under instrumental reason, may come into existence in a communicatively rational manner.

Theorists, such as Innes (1996), Forester (1993), and Duane (1997) have seized upon the notion of communicative action and communicative rationality to develop a theoretical backing and justification for applied collaborative planning. Indeed, the links are clearly present when glancing at the general goals of collaborative planning. In a collaborative process time, energy and resources are devoted to soliciting the needs and concerns of affected parties in a fair open dialogue where creative solutions can be explored and evaluated (Wondolleck and Yaffe, 2000). Opponents and interested parties are brought together to build a common understanding of a situation, develop and test ideas, and design pragmatic solutions. Collaboration implies a joint decision-making approach to problem resolution where power is shared and stakeholders take collective actions and subsequent outcomes from those actions (Selin and Chavez, 1995). Its major characteristics are the following: 1) voluntary participation; 2) direct face to face group interaction among representatives and parties; and 3) mutual agreement or consensus decisions by the parties on the process to be used and the settlement that emerges (Susskind and Cruikshank, 1987; Crowfoot and Wondolleck, 1990; Bingham, 1986).

Conflict Management and Alternative Dispute Resolution

The field of collaborative planning provides insights into ecosystem management in several ways. Managing ecosystems that extend across different types of land ownership, organizations, and jurisdictions often involve conflict. An environmental planning process, which involves input from numerous interests, is also wrought with conflict. Conflict management and negotiation are thus another set of concepts that help form an understanding of how to plan for ecosystems. Many writers and researchers build on the dispute resolution literature to develop a consensus building process (Susskind et al., 1999; Patterson, 1999). This literature began with a focus on negotiation where methods were developed to succeed at distributive bargaining (Bacow and Wheeler, 1984; Carpenter and Kennedy, 1988). More recently the field has evolved to embrace more joint decision-making processes that include concepts of "principled negotiation" (Fisher and Ury, 1991; Godschalk, 1994), where dialogue, information sharing and open communication, the ability to frame and reframe problems, and active listening are central to overcoming inevitable conflicts when dealing with multiple interests (Selin and Chavez, 1995; Daniels and Walker 1996; Daniels et al., 1996).

The field also offers specific tools and techniques to facilitate conflict management, such as appropriate forums, use of a single text during negotiations (this would be the plan in this study), and the use of a third party as a mediator or facilitator (Susskind and Cruikshank, 1987; Crowfoot and Wondolleck, 1990; Godschalk, 1992; McGinnis et al., 1999). While it is beyond the scope of this chapter to discuss all of the benefits of these techniques, it is important to understand that they provide a "how to" level of detail when bringing parties together to work on common natural resource problems. This knowledge is essential when planning for ecosystems since, as described above, it will inevitably involve multiple, and sometimes competing interests.

Since ecosystem management and planning is, in many ways, an exercise in conflict management, relying on its tools is an important part of developing a plan that seeks to manage natural systems. Conflict management techniques, such as information sharing, joint database production, and specific alternative dispute processes contribute to an understanding of how to effectively plan for ecosystems by offering tools that enable multiple interests to reach a decision about what their communities will look like in the future.

Informal Relationships and Social Capital

While social capital and the role of norms in collaborative planning are largely borrowed concepts (taken, for example, from Putnam, 1993), they are discussed frequently in the collaborative planning literature and are critical components in understanding collaborative ecosystem planning. Voluntary cooperation among participants is enhanced when there is a substantial stock of social capital in the form of norms (Ostrom, 1990; Innes, 1996; Duane, 1997). Norms arise out of personal relationships, which develop independent of formal structures. Norms help explain the reliance on collective action at time when it would seem that individuals would behave more selfishly to further their interests. Reciprocity is the most important norm in facilitating collective action because it helps create mutual trust and reputation (Wondolleck and Yaffee, 2000). Trustworthy individuals who have faith in others with a reputation for being trustworthy can engage in mutually productive social exchanges. There are links between the levels of trust individuals have in others, the investment others make in trustworthy reputations, and the probability that participants will use reciprocity norms (Ostrom, 1990). Without reciprocity, the system of cooperation will break down.

By providing a foundation for the development of informal channels, norms reduce conflicts within and between organizations; they help build the type of human relationships needed to collaborate across multiple jurisdictions, government agencies, and resource users. Critical to the maintenance of this collaboration norm is face-to-face communication. Horizontal communication and information networks allow individuals to develop trust in the reliability of others, increasing the chances that they will reciprocate with trust and cooperation. Again, by focusing on the people and informal relationships, the collaborative planning literature provides a key insight into how to effectively manage transboundary ecological systems. Norms, such as trust and reciprocity help facilitate communication, information sharing, and other types of collaboration among stakeholders, which are essential aspects of ecosystem planning.

In summary, the theory and practice of collaborative planning makes important contributions to understanding how to manage ecological systems. By viewing decision-making as a conflict management process where human values are often driving the discussions, the literature brings important tools and concepts to accomplishing the human side of ecosystem management. Information sharing, joint database production, joint accountability, and specific conflict management processes are all important aspects to managing systems that infringe upon the interests of numerous parties. A locally based ecosystem management plan should not only lay-out specific conflict management processes, but express how it will foster human relationships and collaboration over the long term (for more details see Chapter 4).

While the literature on collaborative planning is strong on people and process, it has several gaps that deter from its ability to inform a definition of sound ecosystem planning. Principally, many of the collaborative process models offered are essentially linear in nature (as offered by Susskind and Cruikshank, 1987; Godschalk et al., 1994; Patterson, 1999; Susskind et al., 1999). Because ecosystems, socioeconomic conditions, and problem typologies constantly change over time, collaborative processes must be considered dynamic and nonlinear. In this sense, managing ecosystems cannot be equated to resolving single disputes at specific time periods. Conflicts are instead multiple, constant, and ongoing (which makes ecosystem planning so elusive). Only an iterative collaborative process that continually revisits conflicts and resource-related problems over long periods of time can adequately support ecosystem planning. The plan, in this sense, reflects a community story with no ending, a set of problems with no single solution. Finally, while collaborative planning provides sound processes and collaborative principles, it does not generate clear goals for ecological management. Better incorporation of ecological principles is necessary. Since most planners cannot become natural scientists, they must have a better understanding of ecological principles, particularly when dealing with the maintenance or restoration of natural systems.

Summary

After a brief overview of ecosystem management as an emerging paradigm for environmental policy in the US, this chapter looks more closely at the theoretical traditions underlying a conceptual definition of ecosystem plan quality. Three major literatures were examined, each with their own contributions to forming an understanding of how to manage ecological systems over the long term. First, the theory on ecosystem science and management provides natural science principles for understanding how ecosystems behave. When drafting policies, planners must be aware that ecosystems operate on broad spatial and temporal scales, function through ecological structure, diversity, and overall integrity, and constantly change over time. Second, the organizational design literature offers a social science perspective based on collaboration across political and jurisdictional boundaries. Ecosystem management must accommodate the fact that ecosystems often span multiple lines of ownership and authority. The theory on inter-organizational coordination and systems thinking provides insights into how organizations can coordinate their efforts to manage what are often times broad ecological systems. Finally, the literature on collaborative planning contributes to an understanding of effective ecosystem management by focusing on the people and the process of making sound decisions. Theories on collaboration, conflict management, and social capital add yet another pillar in laying a foundation for understanding how to plan at the ecosystem level. Together, these three areas of literature provide the theories and concepts that lead us to a definition of ecosystem plan quality presented in Chapter 4.

Chapter 3

Principles of Effective Ecosystem Management and Planning

While the review of three literatures in the previous chapter covers a great deal of theoretical ground, it is still necessary to lay a more directed foundation for how to effectively plan at the ecosystem level. To make sense of the multiple concepts associated with ecosystem approaches to management, this chapter synthesizes the contributions of each literature into a set of principles for effective ecosystem planning. These principles take the essential elements of each literature area previously described and provide a more coherent framework for understanding ecosystem approaches to management. They are not meant to be testable hypotheses, but rather guide the reader in understanding what makes for effective ecosystem management, particularly at the local level. In this way, the principles act as the next step in forming a definition of effective local ecosystem planning. These principles will be either directly or indirectly captured in the definition of local ecosystem plan quality developed in Chapter 4. Plans that clearly articulate the concepts behind sound transboundary resource management will more likely lead to collective actions that protect ecological functions while meeting the interests of human communities.

I. Protecting Regionally Significant Habitats: By focusing protection efforts on patches and corridors that serve as important stepping stones or integral ecological components of a larger natural system, the goals of ecosystem management (protecting integrity through protecting biodiversity) will be better attained. Regionally significant habitats comprise the landscape mosaic, which is an essential feature of ecosystem protection. Protecting regionally significant habitats will facilitate the effective management of ecological systems by maintaining levels of overall biodiversity, and the structure, function and integrity of natural systems (see Chapter 6 for more detail). This principle stems directly from the literature of landscape ecology and conservation biology.

II. Developing a Sense of Place: Sense of place is defined by the collection of meanings, beliefs, and feelings individuals or groups associate with a particular locality (Williams and Stewart, 1998). Ecosystem management is about developing a sense of place because it relies on human values to determine the desired future state of a landscape (Wondolleck and Yaffee, 2000). As Power (1996) remarks, the fact that people care where they live instills a sense of place that is valued over time, not only by rational economic theory, but by the emotional attachment people form and associate with a high quality of life. Scientific information is an important part of forming these values, but is not an end in itself. Sense of place becomes a valuable concept in management because it can help reframe an individual's perception of management space. It creates a sense of purpose that brings stakeholders together to collaboratively plan for the future (McGinnis et al., 1999). Rather than concentrate on a narrowly defined area, such as a single park or local community, stakeholders can focus on a broader sense of place to reorient their perspective to include ecological functions and processes. In other words, they can better think about and act upon an ecosystem, as opposed to a single jurisdiction.

Sense of place is a starting principle for ecosystem management because it emphasizes the human component in addressing transboundary resource issues. It helps us realize that ecosystem management is not a technical problem, but a people problem solved by managing people to reach a collective vision of resource use and protection. By developing a sense of place, it forces us to understand and act upon the holistic nature of ecosystem management by directly incorporating human values. In this sense, the concept acts as a bridging mechanism between the science of ecosystems and their management. Sense of place also allows us to better reconcile wide ranging spatial and temporal scales essential to effective ecosystem management practices. Managers are no longer tied to the boundaries of a single resource or a time frame based on the cycle of political elections. Instead, the management space can be easily extended to include overlapping ecological processes, communities of interest, and time frames measured in generations. It is only by developing a strong and enduring sense of place that we will be able to craft plans that seek to attain a collective vision of how to harmonize with the natural world.

III. Incorporating Systems Thinking: Systems thinking principles apply to both understanding the behavior of natural systems and creating organizations capable of making ecosystem level decisions. It is the integration of these two areas that enables effective ecosystem management. Systems thinking relies upon principles of ecology to explain organizational behavior and reveal paths for organizational change. Feedback, dynamic complexity, and interrelationships are concepts that scientists use to understand and predict workings of ecological systems (Wheatley, 1992). By applying these concepts to organizations, systems thinking provides insight into how to actually manage ecological systems by essentially mimicking their behavior. Furthermore, the system (both ecological and organizational) is considered open, where energy and information is constantly exchanged with outside environments. Thus, a systems focus requires cross-jurisdictional problem solving where issues extend across traditional interests and coalitions (Yaffee, 1996). Such an approach is needed to manage dynamic ecosystems over long timeframes, involving numerous parties and concerns, where information is never complete and the ecological and sociopolitical conditions are always in flux.

Management that focuses on larger spatial or temporal scales, or that highlights interconnections necessarily involves a greater level of interrelationships among stakeholders and landowners (Wondolleck and Yaffee, 2000). Systems thinking will help organizations learn and adapt as they manage the environment. This approach acts as the basis for collaborative learning, which emphasizes public involvement, joint learning, open communication, and appropriate change (Daniels and Walker, 1996). Incorporating systems thinking into organizational behavior will help organizations overcome political, institutional, and jurisdictional boundaries necessary to for effective ecosystem management. Because collaborative ecosystem management is ongoing and concerned with large temporal scales, coordination and conflict management involves continual iterations. Collaboration is not fixed in time, but participants and organizations learn as they move toward managing ecological systems. Thus, successful ecosystem management distinguishes itself by fostering a collaborative environmental learning (CEL) environment (Randolph and Bauer, 1999; Cortner and Moote, 1999).

IV. A Proactive Approach to Planning and Management: While it is human nature to act on the basis of a perceived threat or impending disaster, a proactive approach to planning and management is more effective in protecting ecosystem processes over the long term. Establishing a protective framework in the beginning stages of decline or when future adverse environmental impacts are anticipated enables managers to more effectively maintain the value (ecological, economic, aesthetic, etc.) of natural resources and overall integrity of ecological systems. Ecosystem-level goals and policies that seek to protect critical habitats and ecological functions in advance, rather than restore them at a later stage (as most do), will save money in the long run and reduce the likelihood of user conflicts or intense competition over scarce remaining resources. Early protection measures are thus more effective in meeting the needs of a diversity of stakeholders and resource users. A proactive approach to planning will facilitate protection of biodiversity before it is diminished and allow managers to protect ecological systems over the long term.

A proactive approach to ecosystem planning and management also allows policy makers to employ the "precautionary principle" (Agardy, 1994). Protecting certain areas before environmental degradation occurs can help to create a buffer against unforeseen, yet potentially devastating events or management mistakes. These protected areas act as an insurance policy, allowing managers to make conservative decisions in the face of scientific uncertainty and increasing environmental decline. Complete ecological understanding or baseline data is thus not a mandatory condition for implementing an ecosystem policy framework. In fact, the precautionary principle is often applied in response to lack of scientific knowledge or full comprehension of the effects of human activities on ecological systems. Management decisions should be based on accurate data and sound ecological knowledge. Hasty initiatives without scientific basis would be a mistake. However, the best plans are always devised under uncertainty and in order to take a proactive approach to management, there must be some estimation or intuitive leaps made in implementing ecosystem policies.

V. Practicing Adaptive Management: One of the most important goals for organizations implementing ecosystem policies is to manage adaptively. As mentioned in the previous chapter, managers must be able to react to constantly changing ecological systems, sudden shifts in interests and objectives, and a continuous barrage of new and often ambiguous information. Ecosystem-based policies and actions need to become flexible instruments, geared for uncertainty and surprise. Adaptive management is an evolving concept where policies are designed as hypotheses and management implemented as experiments to test those hypotheses (Lee, 1993; Holling, 1978). In most cases, hypotheses are predictions about how one or more species will respond to management actions. The rule of good experimentation, however, is that the consequences of the actions be potentially reversible and that the experimenter learns from the experiment (Holling, 1996). The concept of adaptive management is borrowed largely from fisheries ecology and management. Fisheries managers test the impacts of regulations by modeling the behavior of fish stocks and then adjust regulations based on achieving a desired effect. This concept has been taken up by Lee (1993) and others and has been applied to more general resource management situations.

In its broadest sense, adaptive management ensures that organizations responsible for setting ecosystem policies are responsive to the variations, rhythms, and cycles of change in the system (both ecological and human) and are able to react quickly with appropriate management techniques (Westley, 1995). The process is relatively straightforward: new information is identified, evaluated, and a determination is made whether to adjust strategy or goals (Lessard, 1998). Adaptive management is a continuous process of action-based planning, monitoring, researching and adjusting with the objective of improving future management actions (Endter-Wada et al., 1998; Holling, 1995). Monitoring is an essential activity for the learning organization and is a central component of a high quality plan. There must be built-in mechanisms to assess how well not one, but several organizations are managing an ecological system. Ongoing monitoring and assessment provide feedback to ecosystem management participants so they can manage adaptively over time. They also instill accountability, particularly if the monitoring party is a participant in the ecosystem management initiative (Ostrom, 1990).

VI. Inter-organizational Collaboration and Capabilities within Ecological Systems: Ecosystem management calls for management across ecological, political, generational, and ownership boundaries. When management units are defined ecologically rather than politically, greater coordination among local landowners and between private landowners and natural resource management agencies is required. All parties must make management decisions collectively because in most cases no single entity has jurisdiction over all aspects of an ecosystem. Thus, ecosystem management requires the active support of a broad cross section of society. The need to integrate the values and knowledge of a broad array of organizations and individuals implies a need to blend organizational and community planning through collaboration among resource owners, managers, and users (Cortner and Moote, 1999). A greater degree of inter-organizational collaboration and capabilities will increase the effectiveness of ecosystem approaches to management.

Building a shared vision across multiple organizations and jurisdictional lines is a difficult task, particularly when each party maintains its own set of interests. Many successful ecosystem management efforts have relied upon conflict management practices to enable multiple individuals and groups to work together towards a common protection goal. While ecosystem management may rely in part on scientific understanding, bringing a diverse and often competing set of interests together to solve transboundary resource problems can define its implementation.

It therefore involves managing human conflict as much if not more, than managing critical natural resources.

VII. Building Informal Relationships: Collaboration at the ecosystem level will depend on the development and maintenance of informal relationships and information channels (Yaffe and Wondolleck, 1997; Randolph and Bauer, 1999; Wondolleck and Yaffe, 2000). Personal relationships and networks among organizations stress communication and increase informal ties that become catalysts for coordination (McGinnis et al., 1999). Because these relationships often supersede rank or position, they allow information to flow more freely and facilitate creative action that would otherwise not take place under more formal structures. Informal mechanisms directly promote coordination and collective action through socialization, knowledge acquisition, and the development of extra-organizational loyalties (Chisholm, 1989). As mentioned in the previous chapter, informal relationships build social capital necessary for maintaining relationships over the long term (Innes, 1996). Social capital facilitates an exchange of information and horizontal communication that is necessary for collaborative ecosystem management. The development of informal relationships will facilitate effective ecosystem management by enabling parties to trust each other, share information, and engage in reciprocal exchanges. In other words, it helps them collaborate more effectively.

Without the appropriate setting, organizational structure or climate, norms and communication may never develop. Instead, individuals will be forced into more rational models of behavior where coordination is superseded by self-interest. Norms are not only learned, but also cultivated through, what is in many ways, a cultural process within communities and organizations. Their development not only depends on the appropriate structure or setting, but on the history of the organization and the relationships within them. If past traditions promote the development of trust, communication, and reciprocity, coordination will most likely occur. On the other hand, if some historic event creates patterns of distrust and malevolent feelings, norms will be slow to develop if they do so at all. In this sense, organizations must be treated as individual cultures comprised of unique sets of norms, rituals, and assumptions. The key to collaborative ecosystem management is to facilitate understanding and integration of various organizational cultures so that participants can more effectively work towards a collective vision of resource protection.

VIII. Sharing Power and Information: Regardless of the type of collaborative arrangement or parties involved, there are several factors that help foster cooperation among different organizations and individuals. For example, potential partners must share in defining the problem. Participation builds ownership over a process to solve a particular problem and increases willingness to cooperate with others. The sharing of power is another important practice that can facilitate collaboration. Asymmetric power distribution can deter the formation of partnerships because no party wants to enter into an agreement on unequal footing. Because potential partners never come to the table with equal power, the challenge is to encourage participants to relinquish and share power (Lowry et al., 1997).

Collaboration among government agencies, organizations, and industries is essential to achieving effective ecosystem approaches to management. However, parties will be less likely to enter into joint problem solving and integrative bargaining arrangements as a result of top-down directives. In ecosystem management situations, where no one is clearly in charge, participants must be persuaded or inspired to coordinate with others and take collective action. There must be strong motivation to share power when it is traditionally hoarded and to take the time and energy to make decisions for the common good. Ostrom (1990) argues that parties will collaborate when perceived net benefits exceed net costs. These realizations are not always readily apparent, which is why leaders and their ability to instill motivation in others are so important. Participants contemplating ecosystem management projects will collaborate only when they understand that doing so will enable them to better meet overarching goals. They must realize that pooling resources is more efficient and including parties rather than excluding them will reduce controversy, allowing the policy-making process to run more smoothly over the long term.

In reality, information is power and the way it is collected, stored, and disseminated is a crucial part of designing effective approaches to ecosystem management (Grumbine, 1994). Information is not simply a neutral commodity passed back and forth in a rational system to make wise management decisions. It is created and looked upon through a system of human values and becomes a vehicle for expressing the way we choose to view the natural world. One of the largest barriers to ecosystem management is not acquiring enough information, but sharing it across jurisdictional boundaries, agencies, and other organizations (Lee, 1992). To this end, data must be widely accessible and highly integrated into all stages of the decision-making process. Joint fact finding, information networks, data negotiation, and communication can all help make certain that information (from both the social and natural sciences) critical to understanding ecosystem level issues reaches all of the parties involved (Yaffee and Wondolleck, 1997).

Sharing information means sharing power (Wheatley, 1992). It can help level the organizational playing field, encourage multiple parties to become more involved in meaningful decisions, and enable those at every level to see the "big picture" (McGinnis et al., 1999). Access to information enables interested parties to secure a seat at the decision-making table and to become active participants in ecosystem management initiatives. In general, sharing information and power will result in a high quality management plan and more effective management of ecological systems. Information viewed as a management tool can be incorporated into the policy process in many ways. One of the most promising tools for utilizing data at the ecosystem-level is Geographic Information Systems (GIS). GIS enables managers and the general public to store, analyze, and graphically depict information associated with large scale ecosystems.

Perhaps the most important factor in developing and using information is to build solid relationships between managers, scientists, NGOs, and the general public. Informal relationships ensure a steady, horizontal flow of information across various organizations and individuals. Because these relationships are often independent of rank or position, they work to avoid data pooling at the upper echelons of management structures or information traveling strictly in a vertical fashion through an organization. Wheatley (1992) provides insights into the use of information in ecosystem management. She views organizations as open systems interacting with the world to such an extent that they are constantly changing and renewing themselves over time. Information is the blood coursing though the system, providing life, activity, and innovation. It not only gives life within the organization, but also establishes links to others. Information should not be controlled and does not behave in a linear fashion as some would like to think. Instead information flows in a circular, decentralized pattern to bring people together in different ways so they may find new solutions to old problems. Sharing information in this way can facilitate collaborative actions to understand and protect ecological systems over the long term.

IX. Focused Education and Training: One of the most important principles leading to effective ecosystem management approaches is the use of education and training programs. While collaborative decision-making processes can help build understanding and a sense of ownership, learning is the most profound way to change behavior and create policy change. Education can help re-frame the way individuals view the natural world and provide them with the skills to solve transboundary resource problems.

Because there are so many different types of ecosystem management participants, education must occur at various levels. Interdisciplinary training programs designed to give practicing resource management professionals exposure to the concepts and methods applicable to ecosystem management will increase the effectiveness of government agencies in addressing their newfound goals. Over the long term, academic programs need to change their focus away from traditional disciplinary specialization and toward broad based interdisciplinary learning that enables social and natural scientists to better communicate and collaborate with each other in pursuit of integrated ecosystem analysis (Endter-Wada et al., 1998). To this end, universities should revise their curriculum to include courses in ecological restoration, landscape ecology, conservation biology, and social ecology (Beatley, 2000). Most importantly, students must be taught to understand the interconnectiveness of ecological and human systems and realize that humans are part of the landscape, not necessarily agents of destructive change (Kessler and Salwasser, 1995).

Finally, education should be aimed at the general public and other key stakeholders. Outreach programs will build public awareness on the importance of protecting the value of critical natural resources and maintaining ecological integrity. For example, once a farmer understands why he or she should alter pesticide application techniques to protect a butterfly or habitat in an estuary fifty miles away, the farmer may be more willing to accept new policies and regulations. The most effective way to change attitudes and beliefs is through public involvement. Education and training programs should not only provide the technical skills needed to model and comprehend ecosystem processes, but also allow all parties to appreciate and act upon the system as a whole.

Chapter 4

Developing a Local Ecosystem Plan Coding Protocol and Measurement Framework

Now that the concepts and principles of ecosystem approaches to management have been identified, we need to understand how they can be incorporated into a local plan. The goal of this chapter is to set-forth a standardized protocol defining a high quality ecosystem plan that combines the concepts associated with ecosystem management and plan quality. In this sense, existing conceptions of what makes a high quality plan are extended by integrating the principles of ecosystem approaches to management. Developing a plan evaluation instrument that ties together ecosystem management principles and the theory on plan quality helps address an all important question that serves as the backbone of this book: *what are the main components of or best practices for a sound ecosystem management plan at the local level?* Answering this question moves us from theory to practice in terms of how to effectively manage ecological systems through local level planning tools.

Plans and Plan Quality

In the US, a plan is usually a written expression of a collaborative process (Kaiser and Godschalk, 2000). It is the blueprint for a community's future, a starting point for transforming collective knowledge to action (Baer, 1997). A plan can thus embody the principles of ecosystem management and provide direction for their implementation. It is the jumping-off point for regulations that will protect landscapes and ecological processes for future generations. Most importantly, plans, planning tools, and plan quality provide a directed, measurable approach for ecosystem management. Understanding what makes a high quality plan is the next step in developing a protocol for ecosystem plan quality.

The notion that a plan can indicate both the quality of the planning process and the strength of implementation has emerged in recent years (Talen, 1996; Hoch, 1998). Baer (1997) sets forth a conceptual model for what he calls "plan evaluation" and identifies a set of criteria with which plans can be evaluated. He focuses on plans as a product or outcome of the planning process, as well as a blueprint for future actions. Assessing plan quality involves comparing plans across different localities (comparative quality), asking whether the plan's policies appear to correspond to and advance the articulated plan goals (internal quality), and/or asking to what extent plan policies correspond to independent criteria, such as how well they advance sustainable development or water resources protection.

Chapin and Kaiser (1979; and Kaiser, Godschalk, and Chapin, 1995) first identified the core characteristics of plan quality: fact base, goals, and policies. A strong factual basis, clearly articulated goals, and appropriately directed polices are considered the central elements of a high quality plan. Specifically, fact base refers to the existing local conditions and identifies the needs related to community physical development. Goals represent aspirations, problem abatement, and needs that are premised on shared values. Finally, policies serve as a general guide to decisions (or actions) about the location and type of development to assure that plan goals are achieved (Berke and French, 1994). These plan components can be measured through a series of indicators or issues, allowing for quantitative assessment and analysis of plan quality.

Subsequent empirical studies have applied the core characteristics of plan quality primarily to natural hazard mitigation. Burby et al. (1997) studied local efforts to plan for and mitigate natural hazards in five states: North Carolina, Florida, California, Texas, and Washington. The study used the planning characteristics to determine if state mandates have an influence on plan quality. This work encouraged additional articles that focused on the link between mandates and the quality of local plans (Burby and Dalton, 1994; Berke and French, 1994; Berke et al., 1996; Burby et al., 1997, among others). These articles made important developments in understanding how to conceptualize and measure plan quality in a local comprehensive plan.

In a recent study of hazard mitigation at the state level, Godschalk et al. (1999) expanded the fact base-goals-policies calculus to other plan components such as strategies/actions and implementation. Studies have also applied the core characteristics of plan quality to environmental issues. Berke et al. (1998) examined the quality of plans adopted under New Zealand's Resource Management Act. Berke and Manta (2000) took another important theoretical step by linking plan quality and the principles of sustainable development. Finally, Brody (2003a) and Brody et al. (2003a, 2004) applied plan quality directly to ecosystem approaches to management.

The studies mentioned above not only helped form an understanding of how to measure plan quality, but also yielded insights into the influences on plan quality, particularly from a contextual perspective. For example, Berke et al. (1996) examined the positive influence of wealth and commitment on plan quality associated with natural hazards. Jurisdictions with wealthier population usually have more financial resources to devote to planning staffs and plan development. Residents with high incomes also are often more educated and have more time and interest in participating in the planning process, particularly when it comes to environmental issues. Brody (2003b) found that higher population levels increased the quality of local plans to manage ecological systems. Berke et al. (1998) found that population growth (as a proxy for growth pressure) increased the quality of environmental plan.

In general, jurisdictions with larger populations usually have more complex environmental problems that result in a need for strong planning. Growth pressures are associated with higher levels of disturbance to habitat, resulting in a greater perceived need to protect remaining areas of biodiversity. Furthermore, high population areas tend to have more financial resources and expertise to devote to plan development. Finally, Burby and May (1998) examined the significance of planning capacity as a contextual control variable in a study on plan quality associated with natural hazards. Planning capacity refers to the number of planners that contributed to the development of the comprehensive plan. The higher the planning capacity for a given jurisdiction, the more technical expertise and personnel devoted to producing the plan.

Plan quality is increasingly being used both as an outcome variable for assessing the planning process and as a causal variable for assessing the plan implementation process (see Brody et al., 2005). The ability to code and measure indicators within a plan has made it a widely used instrument with which to quantitatively assess the quality of management efforts. While previous research provides a conceptual and methodological basis for determining the quality of a plan, no study to date has linked plan quality to achieving the principles of ecosystem management. Furthermore, no scholarly work has thoroughly explored how its ecological and social components can be captured and measured in a local land use plan.

Why Comprehensive Plans?

The comprehensive land use plan is proposed as an ideal policy instrument that can encapsulate and implement the major principles of ecosystem management at the local level. The plan, in this case, acts as an ideal regulatory vehicle for realizing these principles and managing ecological systems over the long term. While comprehensive plans are limited to single jurisdictions and are not traditional ecosystem management plans per se, they provide an ideal measure for ecosystem management capabilities at the local level.

First, because these types of plans in Florida need to look beyond jurisdictional boundaries, drive collaborative efforts with other jurisdictions or organizations, and contain policies that seek to protect critical habitats comprising broader ecosystems, they act as strong indicators of how well local jurisdictions will manage ecosystems over the long term. A plan's content and policies often determine a local jurisdiction's level of natural resource use, participation in regional/ecosystem planning efforts, and ability to protect critical natural habitat essential to maintaining ecosystem services. Second, since comprehensive plans are essentially guides to future actions, they take a long-range approach suitable for dealing with temporal scales related to ecosystems. Finally, comprehensive plans are continually being updated to reflect new information and shifts in the public interest. Adaptability is an essential component to address constantly changing ecological and social conditions. A comprehensive plan thus contains all of the characteristics of a traditional ecosystem management plan, only it is focused on the local level.

A high quality ecosystem management plan captures all of the principles and themes comprising ecosystem management and pulls them together as an integrated whole. In a plan, competing values, goals, and views of the natural world are brought together and bound into one document. It represents the end point of a conflict management process, where parties have been able to form a collaborative vision of how to protect ecosystem values into the future. Plans set forth policies, measurable objectives, standards and criteria, and identify who will carry out the proposed solutions. Finally, they set the stage for collective action and innovative change for the common good.

The Protocol

The first major step in developing a protocol for ecosystem plan quality is to extend established planning theory and practice by adding ecosystem considerations to existing conceptions of what constitutes a high quality plan. This protocol builds on and extends previous conceptions of plan quality, which identify factual basis, goals, and policies as its core components (Kaiser et al., 1995) by adding the two additional plan components of inter-organizational coordination and capabilities and implementation. The first additional component captures more accurately the aspects of collaboration and conflict management often required with ecosystem approaches to management. The implementation component measures how likely the goals, objectives, and policies in the plan are to be put in place (not if implementation actually occurred). This component captures, among other issues, the concepts of ecological monitoring, enforcement, and a commitment to put the adopted plan in place. The addition of these components to original conceptions enables the definition of plan quality to more effectively capture the principles of ecosystem management. Plan quality is thus conceptualized (and measured) through the following five components: Factual Basis; Goals and Objectives; Inter-organizational Coordination and Capabilities; Policies, Tools and Strategies; and Implementation.

Together these five plan components constitute the ability of a local plan to manage and protect the integrity of ecological systems. As mentioned above, the five plan components by themselves constitute the basis of a high quality plan but have never been considered with respect to ecosystem approaches to management. Indicators (or issues) within each plan component further "unpack" the conceptions of plan quality. The remainder of this chapter will describe each plan component and its indicators, and briefly show the links to the literature described in earlier sections. A planning protocol listing each plan component and indicator is provided in Table 4.1. A total of 123 indicators within the plan components help operationalize and measure the degree to which local comprehensive plans in Florida are managing natural systems traversing multiple jurisdictions.

Table 4.1 Ecosystem plan coding protocol

Factual basis		
A. Resource inventory		
Ecosystem boundaries/edges	Ecological zones/habitat types	Ecological functions
Species ranges	Habitat corridors	Distributions of vertebrate species
Areas with high biodiversity/ species richness	Vegetation classified	Wildlife classified
Vegetation cover mapped	Threatened and endangered species	Invasive/exotic species
Indicator/keystone species	Soils classified	Wetlands mapped
Climate described	Other water resources	Surface hydrology
Marine resources	Graphic representation of transboundary resources	Other prominent landscapes
B. Ownership patterns		
Conservation lands mapped	Management status identified for conservation lands	Network of conservation lands mapped
Distribution of species within network of conservation lands		
C. Human impacts		
Population growth	Road density	Fragmentation of habitat
Wetlands development	Nutrient loading	Water pollution
Loss of fisheries/marine habitat	Alteration of waterways	Other factors/impacts
Value of biodiversity identified	Existing environmental regulations described	Carrying capacity measured
Incorporation of gap analysis data	Loss of key species	Loss of native vegetation
Boating impacts		
Goals and objectives		
Protect integrity of ecosystem	Protect natural processes/functions	Protect high biodiversity
Maintain intact patches of native species	Establish priorities for native species/habitat protection	Protect rare/unique landscape elements
Protect rare/endangered species	Maintain connection among wildlife habitats	Represent native species within protected areas
Maintain intergenerational sustainability of ecosystems	Balance human use with maintaining viable wildlife populations	Restore ecosystems/ critical habitat
Other goals to protect ecosystems	Goals are clearly specified	Presence of measurable objectives

Table 4.1 continued

Inter-organization coordination and capabilities for ecosystem management		
Other organizations/ stakeholders identified	Coordination with other organizations/ jurisdictions specified	Coordination with adjacent counties
Coordination with state level organizations	Coordination with federal level	Coordination within jurisdiction specified
Intergovernmental bodies specified	Joint database production	Information sharing
Coordination with private sector	Coordination with water management districts	Participation in ecosystem-based initiatives (i.e. NEP, EMAs)
Links between science and policy specified	Position of jurisdiction within bioregion specified	Intergovernmental agreements
Conflict management processes	Commitment of financial resources	Integration with other plans/ policies in the region
Other forms of coordination		
Policies, tools, and strategies		
A. Regulatory tools		
Resource use restrictions	Density restrictions	Restrictions on native vegetation removal
Removal of exotic/ invasive species	Buffer requirements	Fencing controls
Public or vehicular access restrictions	Phasing of development	Controls on construction
Conservation zones/ overlay districts	Performance zoning	Subdivision standards
Protected areas/sanctuaries	Urban growth boundaries to exclude habitat	Targeted growth away from habitat
Capital improvements programming	Site plan review	Habitat restoration actions
Actions to protect resources in other jurisdictions	Establishment of a network of system of protected areas	Create wildlife corridors
Protect threatened or endangered species	Structural or design solutions to protect habitat	Other regulatory tools
B. Incentive-based tools		
Density bonuses	Clustering away from habitats	Transfer of development rights
Preferential tax treatments	Mitigation banking	Specific mitigation measures to protect habitat
Impact fees to protect habitat	Other incentive-based tools	
C. Land acquisition programs		
Fee simple purchase	Conservation easements	Other land acquisition techniques

Table 4.1 continued

D. Other policies, tools, and strategies		
Designation of special taxing districts for acquisition funding	Control of public investments and projects	Public education programs
Studies or ecological surveys		
Implementation mechanisms		
Designation of responsibility	Provision of technical assistance	Identification of costs or funding
Provision of sanctions	Clear timetable for implementation	Regular plan updates and assessments
Enforcement specified	Monitoring for plan effectiveness and response to new information	Monitoring of ecological health and human impacts

Factual Basis

In general, the factual basis of a plan refers to an understanding and inventory of existing resource issues, environmental policies, and stakeholders' interests within the ecosystem. It takes both a written and visual form, and serves as the factual and descriptive basis on which policy decisions within the plan are made. The foundation for the factual basis is a resource inventory of critical natural resources, which draws explicitly from the literature on ecosystem science and landscape ecology. The level of understanding displayed of the boundaries and functions of ecological systems is not only essential to physically managing the landscape, but demonstrates the geographic level of focus and sense of place inherent in a community. The level of knowledge associated with the existing resource base and the adverse impacts to these resources indicates how planners and community members relate to and value the natural systems surrounding them. The factual basis also supports and often drives the other components comprising ecosystem plan quality.

Indicators, such as mapping ecosystems and habitat boundaries, describing ecological functions, and being able to classify wildlife and vegetation all contribute to a strong resource inventory. Identifying and protecting regionally significant habitats is another component of the factual basis that extends the landscape design theory of protecting the landscape mosaic. In order to protect the ecological infrastructure of a landscape, planners must first identify critical habitat, areas of high biodiversity, and most importantly identify corridors that facilitate the movements and migration of key species. Protecting systems of habitats is also a crucial part of the human ownership category of the factual basis. To protect new lands, the existing network must first be identified. The resource inventory combined with the human ownership category provide the basis for a gap analysis that can greatly aid planners in generating plans that seek to manage ecological systems. As mentioned above, if a plan is to adequately manage ecological systems, it must contain a sound factual basis that identifies the resources to be protected.

The final category of the factual basis component of a plan is the identification of human impacts or resource problems. A community must thoroughly understand the condition of or adverse impact on its natural resource base. Identifying impacts, such as human population growth, the development of wetlands, and water pollution also measures how well a community has developed its sense of place. This category goes beyond simply identifying existing impacts on resources by indicating how and where the community attaches values to those resources. It also captures systems thinking by demonstrating the relationship and interactions between humans and the natural system as a whole. Overall, by forming a detailed understanding of existing critical habitats and the impacts to these habitats, planners are able to take a more proactive approach to the management of ecosystems. Policies can be adopted to protect biodiversity before adverse impact takes place in the future. Having information on the state of natural resources also helps communities adapt to changing conditions. Without intimate knowledge of the health of existing resources, planners cannot know what is being threatened and how to respond to ensure a more ecologically sustainable approach to development.

Goals and Objectives

Goals and objectives guide the implementation of ecosystem management. They contain both general statements of long-term goals regarding clarity and consistency, as well as specific measurable objectives, such as a 40 percent reduction in nutrient runoff to reduce impacts on an estuarine system. This plan component is perhaps the best measure of the values of a community to protect regionally significant habitats and the integrity of ecological systems. Goals must be clearly specified and objectives must be measurable as to provide benchmarks of success. Well-defined goals generated through strong leadership are more detailed than vague commitments of ecosystem protection. They penetrate into the meaning of ecosystem management derived from ecosystem science by seeking to maintain large intact patches of native species, connections among significant habitats, and intergenerational sustainability of natural systems. Furthermore, they are geared towards protecting both the functionality of the ecosystem, as well as its unique landscapes and rare species.

Systems thinking is integrated into the conception of plan quality by the degree to which goals are aimed at the entire natural system (even beyond the local jurisdictional boundaries as is found in many high quality comprehensive land use plans), rather than simply a small fragment or location. Ecosystem planning goals and objectives also indicate the degree to which a community takes a proactive approach to resource management. Goals that are aimed at protecting habitat rather than restoring them at a later date better capture the overall intent of ecosystem management and the protection of biological diversity. In summary, goals and objectives are not solely driven by the theory of ecosystem science and landscape ecology. They are instead a reflection of a community's values, sense of place, and commitment to taking a systems approach to proactively maintaining the integrity of natural systems for future generations.

Inter-organizational Coordination and Capabilities

It is largely recognized that ecosystem management is a human boundary-spanning problem (Grumbine, 1994; Vogt et al., 1997; Wondolleck and Yaffe, 2000). Ecological systems, particularly large watersheds and estuaries, extend across multiple jurisdictions, making sustainable management of the entire system a difficult prospect (Kirklin, 1995). Because ecosystems do not adhere to what has become a "crazy quilt" of land ownership, organization, and governance, environmental management goals are not being reached and natural systems, such as the Everglades in Florida continue to decline (Light et al., 1995; Daniels et al., 1996). While this natural system is intricately connected over broad spatial and temporal scales, the land use decision framework is limited to local jurisdictions and limited input from regional planning councils. Uncoordinated local land use decisions have cumulative negative impacts on the system as a whole. In other words, the Everglades and its sub-ecological systems are suffering a death from thousands of locally imperceptible, individual development decisions. Collaboration across jurisdictional lines and among multiple organizations thus becomes imperative if approaches to ecosystem management are to be attained (Daniels and Walker, 1996; Randolph and Bauer, 1999).

Inter-organizational Coordination and Capabilities captures the ability of a local jurisdiction to collaborate with neighboring jurisdictions and organizations to manage what are often transboundary natural resources. It represents a key component in defining local ecosystem plan quality because it measures to what degree a local community is able to recognize the transboundary nature of natural systems in Florida and coordinate with other parties both within and outside of its jurisdictional lines. An intergovernmental coordination plan element is required by the state of Florida, but there is wide variation among plans with regard to protecting natural systems. This plan quality component addresses the critical factors necessary to foster collaboration which include, among other indicators, joint fact finding, information sharing, inter-governmental agreements, and integration with other plans in the region (e.g. Ecosystem Management Area plan, National Estuary Program).

The Inter-organizational Coordination and Capabilities plan component draws not only upon the organizational design literature stressing collaboration, but also on the collaborative planning aspects of building informal relationships and sharing power and information. Both single jurisdictional and transboundary coordination is incorporated through indicators such as information sharing, the designation of intergovernmental bodies and agreements, and integration with other plans. Joint database production, cooperative agreements, and the commitment of financial resources indicate that power and information are being shared among individuals and organizations. These indicators are also a sign of trust and reciprocity formed through informal relationships and interpersonal networks. Use of conflict management processes and the identification of important stakeholders demonstrates that the community is actively pursuing a collaborative process through stakeholder participation. In general, since ecosystem management is so dependent on collaboration among individuals and organizations, this plan component is a crucial aspect of defining and measuring plan quality.

Policies, Tools, and Strategies

Policies, Tools and Strategies represent the heart of a plan because they actualize community goals and objectives by setting forth actions to protect critical habitats and related natural systems. Policies draw heavily on the environmental and land-use planning literatures to identify tools that effectively protect ecological systems. This plan component also looks to the landscape ecology literature to incorporate ecological design principles in the designation of specific tools and strategies. Policies include traditional regulatory tools, such as land use or density restrictions, restrictions on native vegetation removal, and buffer requirements. In addition to regulatory approaches, more innovative incentive-based tools are also incorporated into this plan component, such as clustering, density bonuses, transfer of development rights (TDRs), and mitigation banking.

Land acquisition programs are another important category within the plan protocol because it indicates the ability of jurisdictions to fund the purchase of critical habitats and sensitive lands. Florida is a leader in acquisitions efforts across the country (Beatley, 2000). Under its Preservation 2000 Initiative, the state generated $300 million per year for ten years to fund the acquisition of sensitive lands. However, leadership at the state level has not necessarily translated into local initiatives to acquire areas containing critical habitat. Finally, educational efforts on the importance of protecting significant habitats and ecosystems are also important indicators within this plan component. As mentioned in Chapter 2, educational programs are essential for engaging stakeholders in the planning process and in helping generate a plan that is enduring and enforceable in its implementation.

In summary, by combining land use tools with ecosystem science and the principles of landscape ecology, a range of techniques can be generated to protect natural systems. Strategies, such as designating protected areas, phasing of development, and targeted growth areas away from critical natural areas are a few of the ecologically-driven approaches to protecting biodiversity that are seldom found in traditional land use plans (Duerksen et al., 1997; Beatley, 2000). Higher quality plans will have a greater breadth and scope of these policies to reflect the innovation necessary to manage complex ecological systems.

Implementation Mechanisms

The final component of ecosystem plan quality is implementation, which measures the ability of a plan to become an enduring instrument that is carried forth through regulations and collective action. For comprehensive plans to be effective, implementation must be clearly defined and laid-out for all affected parties. Implementation stems from the theoretical concept of collaborative learning and the practice of adaptive management. It is based not only on the ability of a community to implement its plan in a timely fashion, but also to designate responsibility for actions, enforce adopted standards, and sanction those who fail to comply.

The implementation plan component also focuses on monitoring activities, on the success of policies, and the response to scientific information so that a community can adapt to changing conditions by setting updated standards to most effectively obtain stated goals and objectives. In this manner, the implementation plan quality component incorporates the concept of adaptive management, group learning and flexible behavior in managing ecological systems that constantly change over time and space.

Chapter 5

Measuring and Mapping Ecosystem Plan Quality

This chapter moves us from concepts to application by testing the ecosystem plan protocol developed in Chapter 4 through two Florida case studies. The first case examines the ability of local comprehensive plans in Florida to incorporate the principles of ecosystem management. The second case evaluates the collective capabilities of local jurisdictions to manage large transboundary ecological systems in southern Florida. It combines plan evaluation with Geographic Information Systems (GIS) techniques to map, measure, and analyze the existing mosaic of management across selected ecosystems in the southern portion of the State. By putting the protocol to the test, we can begin to answer the question: *which state-mandated comprehensive plans are most geared to ecosystem management and why?*

CASE 1: A Report Card for Ecosystem Management through Local Land Use Planning

This case examines the ability of local comprehensive plans across Florida to implement the principles of ecosystem management. It seeks to understand how comprehensive plans can effectively contribute to the management of ecological systems by systematically evaluating local plans against a conceptual model (developed in Chapter 4) of what makes for a high quality ecosystem plan. By using the protocol to score a sample of plans, we gain insights into how well local communities are managing ecological systems. This analysis essentially gives a detailed report-card on ecosystem management performance and provides direction on how local communities can improve their environmental frameworks.

The Sample

The first step in this case study was to ensure that the sample of plans evaluated were representative of all local plans across the State. The study population was based on all local jurisdictions (cities and counties) in Florida that have completed under the State mandate recent updates of their comprehensive plans. We then selected for analysis a random sample of 30 communities using the following sampling strategy. 1) The sample included only those jurisdictions with a population of 2,500 or more to make certain the sample was not skewed toward small communities (Berke and French, 1994). 2) The sample excluded large cities, such as Miami because these jurisdictions have very different contextual factors that may skew the sample (Berke et al., 1996). 3) The sample used only coastal jurisdictions to maintain a degree of consistency and comparability in terms of the types of ecosystems assessed.

Scoring the Plans

Each indicator in the ecosystem plan protocol was measured on a 0-2 ordinal scale, where 0 is not identified or mentioned, 1 is suggested or identified but not detailed, and 2 is fully detailed or mandatory in the plan. In the factual basis component of the protocol, most items have more than one indicator. For example, habitats can be either mapped, catalogued or both. An item index was created in these cases by taking the total score and dividing it by the number of sub-indicators (i.e. an item that receives a 1 for mapping and 1 for cataloging received an overall issue score of 1). This procedure assured that all plan quality items remained on a 0-2 scale, while at the same time recognizing that a strong fact base relies on both textual and graphic description. Together, these indicators capture the principles of effective ecosystem management and translate them into elements that can be identified, measured, and compared across each plan in the sample.

Once plans were coded using the ecosystem plan protocol (Table 4.1), two types of scores were calculated. First, an overall measure of ecosystem plan quality was derived by creating indices for each plan component and overall plan quality (as done by Berke et al. (1996) and Berke et al. (1998). Indices were constructed for each plan component based on three steps. First, the actual scores for each indicator were summed within each plan component. Second, the sum of the actual scores was divided by the total possible score for each plan component. Third, this fractional score was multiplied by 10, placing each plan component on a 0-10 scale. Adding the scores of each component (factual basis; goals and objectives; inter-organizational coordination and capabilities; policies; and implementation) resulted in a total plan quality score. Thus, the maximum score for each plan is 50.

Second, to further unpack the results from evaluating plans against the planning protocol, we used several additional measures based primarily on the techniques used in Godschalk et al. (1999). These measures look at each issue-based indicator in the protocol from three perspectives: their presence, their quality, and a total quality issue score.

1. Item breadth score = No. of plans that address item/No. plans in sample (0-1 scale)
2. Item quality score = total score of all plans that addressed an item/no. plans that addressed the issue (0-2 scale, converted to 0-1 scale)
3. Total item score = item breadth + item quality (0-2 scale)

This set of scores provided a sharper lens of focus with which to identify in greater detail the ability of local plans to integrate the principles of ecosystem management. Item breadth measures the percentage of the sample that includes an item in the planning protocol. Item quality measures not only if the item was included in the plan, but its level of detail or the strength of a particular policy (mandatory versus suggested). The total item score combines the previous two measures to provide insights into the overall quality of an item. The significance of an item that is not often included in a plan, but is done so with high quality can thus be factored into the overall score of a plan.

Assessing the Plans

Overview of Ecosystem Plan Quality

Results from the first phase of analysis provide an overall assessment of how well local jurisdictions in Florida are incorporating the principles of ecosystem management into their comprehensive plans. As shown in Table 5.1, the mean score for total ecosystem plan quality is 20.62, which on a scale of 0-50, indicates a relatively weak effort to manage ecological systems at the local level. Mean scores for all plan components (scale of 0-10) also register fairly low despite a strong state program on ecosystem management and a clear local planning mandate to protect critical habitats and ecological functions.

The factual basis is the lowest scoring plan component, demonstrating a lack of knowledge regarding the existing level of critical natural resources within a jurisdiction. In contrast, the inter-organizational coordination and capabilities plan component scores fairly high with a mean of over 5.0 (on a scale of 0 to 10). A high score for this component suggests that jurisdictions recognize the transboundary nature of ecosystems and are willing to collaborate with other jurisdictions to manage these natural resources over the long term. The score, however, may simply reflect the fact that a general inter-governmental coordination element is required in all plans. Specific scores for each plan component are discussed in more detail in the subsequent sections.

Table 5.1 Descriptive plan quality scores for each plan component

Plan component[a]	Mean	Standard deviation
Factual basis	2.25	2.03
Goals and objectives	3.63	2.25
Inter-organizational coordination	5.14	1.92
Tools, policies, strategies	4.35	1.57
Implementation	5.00	2.30
Total ecosystem plan quality[b]	20.62	7.76

Source: Adapted from Brody, 2003: 524.

Note: [a] Maximum score by plan component is 10.00; [b] Maximum score for total ecosystem plan quality is 50.00.

Plan Component and Item Scores

Results from the second phase of analysis provide a more detailed examination of local jurisdictions' ability to incorporate the principles of ecosystem management by unpacking the results from the plan coding protocol item by item.

Factual Basis: In the Resource Inventory category (Table 5.2), a relatively low percentage of plans inventory ecosystem boundaries, ecological functions, areas of high biodiversity, or natural resources that extended beyond the local jurisdiction. These issues form the building blocks for identifying and managing ecosystems. Instead, the majority of plans concentrate on traditional environmental components within jurisdictions, such as soil types, wetlands, and surface water features. Other important elements for understanding ecosystem processes, such as identification of species ranges, keystone species, and exotic or invasive species receive some of the lowest scores in terms of breadth. Habitat corridors between wildlands, an essential part of maintaining the landscape mosaic because they allow for natural movements of species, are not mapped or described by any of the plans sampled. Vegetation mapping and classification is more likely to be included over vertebrate species since land cover is more easily identified and modeled graphically across landscapes. Only a few jurisdictions, such as Pinellas County and the city of Bradenton use GIS to generate maps of resources or biodiversity, despite the fact that these data are readily available from the state.

While most of the plans do not tend to focus on ecosystem-based environmental factors, when they do descriptions are done in detail, resulting in high item quality scores (as opposed to overall plan component scores). This result suggests that when local jurisdictions make the commitment to move beyond the standard for inventorying critical natural resources (soils, wetlands, surface water, etc.), they ensure a high quality result. This phenomenon causes the total item scores for ecosystem-based environmental issues to be relatively higher. For example, only just over half of the plans sampled describe the ecological functions for habitat type or ecological zones, but this item receives the second highest total item score (1.41) in the Resource Inventory category. Similarly, only 47 percent of the sample mapped their land cover, but did so with such high quality that the total item scores for this indicator 1.29, ranking it among the highest in its category.

Human impacts listed and described in the sample of plans concentrate primarily on typical urban environmental problems, such as water pollution (63 percent of the sample) and nutrient loading (50 percent). Federal water quality monitoring regulations and obvious environmental disturbances, such as eutrophication easily identify these impacts. In contrast, relatively few plans address the most pertinent issues related to habitat degradation and ecosystem decline in Florida and other states, such as habitat fragmentation, loss of wetlands, or an increase in road density. Experts cite these issues as having the greatest adverse impacts on ecosystems and the decline of biodiversity across the state (Cox et al., 1994; Noss and Cooperrider, 1994; Beatley, 2000). Most items in this category are discussed in detail and receive relatively high quality scores. Scores are of particularly high quality in instances where monitoring programs are in place or information is available at the state level, such as for water pollution and nutrient loading.

Table 5.2 Issue-based scores for the factual basis plan component

Indicator	Issue breadth	Issue quality	Total issue quality
Resource inventory			
Ecosystem boundaries	.33	.53	0.86
Ecological zones/habitats	.67	.66	1.33
Ecological functions	.53	.88	1.41
Species ranges	.23	.50	0.73
Habitat corridors	.00	.00	0.00
Vertebrate species	.17	.80.	0.97
Biodiversity/species richness	.33	.63	0.96
Vegetation classified	.57	.62	1.18
Wildlife classified	.47	.50	0.97
Land cover mapped	.47	.82	1.29
Threatened/endangered species	.53	.52	1.05
Exotic species	.17	.50	0.67
Keystone species	.13	.56	0.70
Soil types/associations	.90	.77	1.67
Wetlands mapped/described	.80	.59	1.39
Climate	.30	.89	1.19
Groundwater resources	.70	.60	1.30
Surface hydrology	.73	.66	1.39
Marine resources	.67	.41	1.08
Representation of transboundary resources	.23	.61	0.84
Other prominent landscapes	.43	.44	0.88
Ownership patterns			
Conservation lands mapped	.43	.38	0.82
Management status for conservation lands identified	.17	.50	0.67
Network of conservation lands mapped	.23	.79	1.02
Distribution of species within network of conservation lands identified	.00	.00	0.00
Human impacts			
Human population growth	.30	.83	1.13
Road density	.03	.50	0.53
Fragmentation of habitat	.23	.71	0.95
Wetlands development	.10	.50	0.60
Nutrient loading	.50	.87	1.37
Water pollution	.63	.87	1.50
Loss of fisheries/marine habitat	.20	.75	0.95
Alteration of waterways	.33	.75	1.08
Other impacts/loss of biodiversity	.63	.79	1.42

Source: Adapted from Brody, 2003: 526.

Goals and Objectives: Table 5.3 reports the number of times a goal or objective in the ecosystem planning protocol is reported by plans in the sample (quality scores were not reported for this plan component to simplify the interpretation of the data). The majority of plans include broad goals to protect the integrity, natural functions, and processes of ecosystems. However, comparatively few plans cite more specific objectives involved in managing ecological systems, such as protecting biodiversity hotspots (23 percent), maintaining large intact patches of native species (37 percent), or maintaining wildlife corridors (27 percent). These results suggest that while plans frequently state general (and often vague) goals related to ecosystem management, they are unable to incorporate specific objectives that could drive precise land use tools and policies.

Protecting rare and endangered species is one of the most frequently stated goals in the sample (80 percent), driven mostly by interest in protecting characteristic megafauna, such as the Manatee (*Trichechus manatus*) or Florida Panther (*Puma concolor coryi*). (Need to provide the scientific names of the manatee and the Florida panther in parentheses following the common names) Planners and planning participants often are well aware of the decline of single species (usually large mammals), but are unable to relate the protection of these species to protecting networks of habitat or areas of high biodiversity. Perhaps this result stems from the historic focus on single species in the United States through the Endangered Species Act (ESA), rather than protecting connected habitats or entire ecosystems. Finally, the majority of plans mention restoration goals and objectives, reflecting the degraded state of many urban areas included in the sample. Most jurisdictions have little remaining viable habitat to protect due to rapid urban development in the 1970s and early 1980s, and instead must focus on goals to restore badly degraded natural systems.

Table 5.3 Issue-based scores for the goals and objectives plan component

Indicator	Issue breadth
Protect ecosystem integrity	.80
Protect natural processes/functions	.83
Protect high biodiversity	.23
Maintain intact patches of native species	.37
Establish priorities for native species/habitat protection	.50
Protect rare/endangered landscape elements	.50
Protect rare/endangered species	.80
Maintain connections among wildlife habitats	.27
Represent native species within protected areas	.10
Maintain intergenerational sustainability of ecosystems	.23
Balance human use with maintenance of viable wildlife populations	.40
Restore ecosystems/critical habitat	.70
Other goals to protect ecosystems	.53
Presence of measurable objectives	.70

Source: Adapted from Brody, 2003: 527.

Inter-organizational Coordination and Capabilities: Overall, results for this category of the planning protocol reveal a strong commitment toward collaborating both within a jurisdiction and with neighboring communities. As shown in Table 5.4, almost all of the jurisdictions sampled mention in detail coordinating with other organizations to protect resources within their boundaries as well as those that cross several administrative lines. Furthermore, most of the jurisdictions (83 percent) express a commitment to integrating other environmental plans or policies in the region into their local planning frameworks. Incorporating regional environmental efforts, such as Water Management District Plans or National Estuary Program plans remains an essential part of achieving ecosystem approaches to management at the local level. Not only do the majority of organizations include language to collaborate to manage ecological systems, but these policies are almost always mandatory, raising their item quality scores.

Item scores are not as strong when it comes to describing the specifics of inter-organizational coordination. Less than half of the sample designates intergovernmental bodies to protect transboundary resources or engage in joint database production. Half of the plans outline conflict management processes to resolve resource conflicts prevalent in ecosystem management. Finally, 20 percent of the plans actually commit financial resources necessary to bring together various parties to manage ecological systems. Although the breadth of these items is low, their item quality is comparatively high. In other words, when an item is included in the plan, jurisdictions generally show a strong commitment to carry it out, which is reflected in the strength of the total item scores.

Table 5.4 Issue-based scores for the inter-organizational coordination and capabilities plan component

Indicator	Issue breadth	Issue quality	Total issue quality
Other organizations/stakeholders identified	.87	.83	1.69
Coordination to protect transboundary resources	1.00	.97	1.97
Coordination within jurisdiction to protect ecosystems	.97	.90	1.86
Intergovernmental bodies specified	.43	.85	1.28
Joint database production specified	.43	.85	1.28
Information sharing	.70	.79	1.49
Links between science and policy identified	.23	.71	.95
Position of jurisdiction within bioregion specified	.43	.65	1.09
Intergovernmental agreements (IGA) designated	.57	.74	1.30
Integration with other environmental plans/policies	.83	.86	1.69
Conflict management process outlined	.50	.87	1.37
Commitment of financial resources	.20	.75	.95
Other forms of coordination	.80	.85	1.65

Source: Adapted from Brody, 2003: 529.

Policies, Tools, and Strategies: Results for this component demonstrate that plans tend to favor traditional environmental policies, such as resource use restrictions in and around critical habitats, restrictions on removal of native vegetation, and conservation zones to protect sensitive lands (Table 5.5). Other regulations, such as fencing controls to permit natural movement of native species (e.g. Florida panther), phasing of development to reduce wildlife disturbance, or setting urban growth boundaries that do not include critical habitats, are less represented. While mainstream policies play an important role in ecosystem approaches to management, the evidence increasingly shows that less commonly used growth management tools focusing on both overall growth patterns (e.g. targeted growth areas) and specific site-related regulations (e.g. subdivision standards) may allow for significant gains in protecting regionally significant habitats (Duerksen et al., 1997). Notably, however, when a policy is stated, it is almost always mandatory, contributing to high item quality scores for indicators within this component. Overall, traditional environmental policies, such as resource use restrictions, native vegetation removal restrictions, and conservation zones, however, receive the highest total item scores in the regulatory category.

Despite their effectiveness in protecting critical habitats and ecological systems (Duerksen et al., 1997; Peck, 1998; Beatley, 2000), incentive-based policies enjoy far less representation than regulatory techniques. The most widely used tool is transfer of development rights (47 percent) to protect primarily wetland habitat. Only 20 percent of the sample cites mitigation banking, despite a strong state-level program and regulatory framework allowing for the practice. When a plan includes incentive-based tools, the policies are almost always mandatory, causing the item quality scores to be extremely high in this section. Low breadth scores account for comparatively low total issue scores for these items.

Seventy-one percent of the sample mentions land acquisition programs, where localities include specific policies to acquire land for conservation to protect critical habitats. This high score might reflect a state level emphasis on the policy, such as the Preservation 2000 initiative, where the state sold bonds sufficient to generate $3 billion over a ten-year period (Beatley, 2000). Land acquisition techniques get incorporated into plans primarily in the form of fee simple purchases.

Other non-regulatory techniques are also important indicators of determining ecosystem plan quality. For example, most plans (87 percent) contain the policy of monitoring ecological processes and human impacts, an essential component of adaptive management. Monitoring policies primarily are associated with water quality issues, but several jurisdictions also include policies for specific species, wetlands habitats, and other ecosystem components. Finally, 50 percent of the plans include educational programs on the importance of protecting habitat and ecological systems. Although the environmental planning arena largely overlooks educating the public, policies can build an understanding of ecological problems and commitment to protecting ecological systems over the long term. When included, polices in this section of the planning protocol are almost always mandatory and the item quality scores thus rate extremely high.

Table 5.5 Issue-based scores for the policies, tools, and plan component

Indicator	Issue breadth	Issue quality	Total issue quality
Regulatory tools			
Resource use restrictions	.83	.96	1.81
Density restrictions	.53	.94	1.47
Restrictions on native vegetation removal	.97	1.00	1.97
Exotic species controls	.60	1.00	1.60
Buffer requirements	.60	.97	1.57
Fencing controls to allow species movement	.10	1.00	1.10
Public or vehicular access controls	.60	.97	1.57
Phasing of development to protect habitat	.03	1.00	1.03
Controls on construction to protect habitat	.93	1.00	1.93
Conservation zones/overlay districts	.87	1.00	1.87
Performance zoning to protect habitat	.20	1.00	1.20
Subdivision standards to protect habitat	.13	1.00	1.13
Protected areas/sanctuaries	.57	1.00	1.57
Urban growth boundaries to protect ecosystems	.03	1.00	1.03
Targeted growth areas to protect habitat	.30	.94	1.24
Capital improvements programming	.27	.94	1.20
Site plan review to protect habitat	.67	.98	1.64
Habitat restoration	.83	1.00	1.83
Actions to protect resources in other jurisdictions	.90	1.00	1.90
Other regulatory tools	.83	1.00	1.83
Incentive-based tools			
Density bonuses	.37	.86	1.23
Clustering development away from critical habitat	.40	.92	1.32
Transfer of development rights	.47	.93	1.40
Preferential tax treatments	.10	.67	.77
Mitigation banking	.20	.92	1.12
Other incentive-based tools	.17	.77	.93
Land Acquisition Programs	.70	.63	1.33
Other policies, tools, and strategies			
Designation of special taxing districts	.07	1.00	1.07
Control of public investments and projects	.53	.94	1.47
Public education programs	.50	1.00	1.50
Monitoring ecological health and human impacts	.87	.79	1.66

Source: Adapted from Brody, 2003: 531.

Implementation: Compared to other plan components, Implementation scores are strong in both breadth and quality (Table 5.6). It is important to note that these results measure a jurisdiction's future ability to implement its plan, not if the plan was actually implemented after adoption. The majority of jurisdictions incorporate the essentials of implementing a plan, such as accountability, a clear timetable, and regular updates or assessments (although one might expect even higher breadth scores given the state mandate to implement a plan). Experts frequently rely upon monitoring plan effectiveness and incorporating new information into updates essential to effective ecosystem management (Lee, 1993). The Implementation component may score relatively high in part due to the high item quality scores in the plans. For example, when a policy is stated, it is almost always mandatory. Jurisdictions do not cite identification of funding for implementation and sanctions for failure to implement policies as frequently as one might expect. These issues, along with enforcement measures, are important because they ensure that policies and projects required in the plan actually come to fruition and are adhered to by the public.

Table 5.6 Issue-based scores for the implementation plan component

Indicator	Issue breadth	Issue quality	Total issue quality
Designation of responsibility	.80	.88	1.68
Provision of technical assistance	.30	.94	1.24
Identification of costs or funding	.33	.85	1.18
Provision of sanctions for failure to comply	.10	1.00	1.10
Clear timetable for implementation	.77	.98	1.74
Regular plan updates and assessments	.67	.98	1.64
Enforcement specified	.67	1.00	1.67
Monitoring for plan effectiveness and response to new information	.77	.75	1.52

Source: Adapted from Brody, 2003: 532.

Assessment of Specific Jurisdictions

Indices for total plan quality were also calculated for each jurisdiction to better understand which specific communities have high or low plan quality. As shown in Figure 5.1, Pinellas County, the city of Jacksonville, and Martin County stand out as the highest scoring plans overall. In contrast, smaller jurisdictions, such as Valpairiso, Miami Shores, and Niceville are among the lowest scoring plans in the sample.

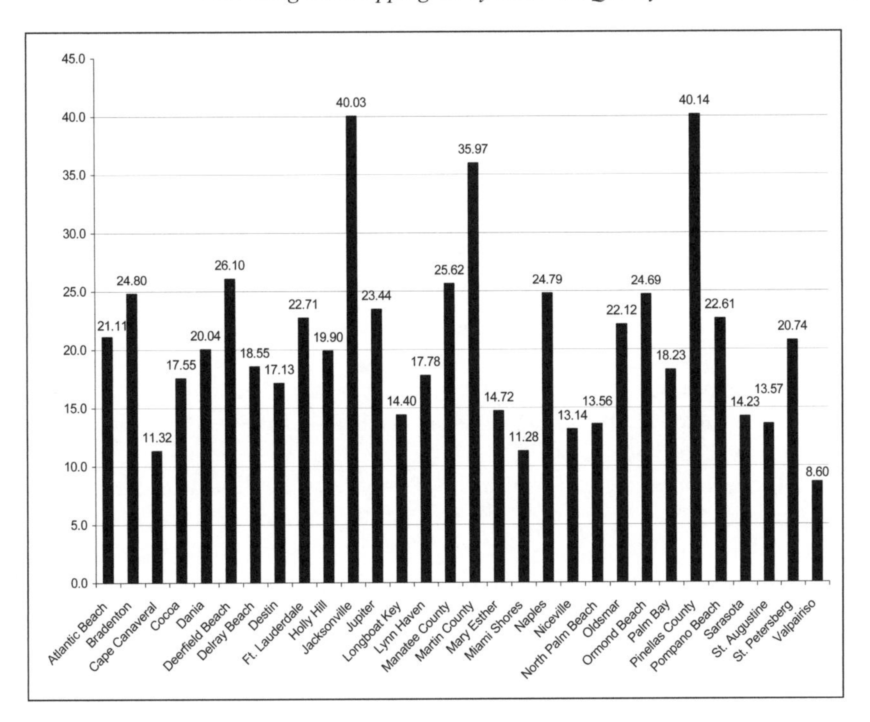

Figure 5.1 Total plan quality

When examining specific plan components for jurisdictions, there appears to be a lack of consistency among scores, where one or two components stand out from the others in terms of influencing total plan quality. In general, strong Inter-organizational Coordination and Capabilities scores across the sample pull total quality higher, particularly in the cases of Atlantic Beach and Mary Esther. High Implementation scores often have the same effect, driving total quality scores higher as is evidenced by Pompano Beach, St. Petersburg, and Ft. Lauderdale. The lack of consistency among plan component scores is an important factor in explaining how a plan, as a total growth management package, actualizes the principles of ecosystem management.

Making the Grade?

Based on the results above, one can infer that local jurisdictions in Florida have not been able to effectively incorporate the principles of ecosystem management into their planning frameworks. While strong interest in ecosystem management exists at the state and regional levels, this commitment has not entirely filtered down to the local level or local jurisdictions have been unable to effectively convert the principles of ecosystem management into their land use planning instruments: comprehensive plans.

Overall, the Factual Basis for the sample lacks detail and fails to address many of the issues associated with managing ecological systems. Some jurisdictions (Longboat Key, Cocoa, etc.) do not even have a factual basis incorporated in their plans and instead rely on separate (often outdated) documents that are neither part of the legal plan nor circulated to the public.

In general, plans reveal a commitment to the broad notions supporting ecosystem protection, but do not include clear and concise goals, which are needed to implement effective ecosystem policies. When goals are stated, they usually are vague and unfocused in their intent. Furthermore, while the majority of plans include measurable objectives to achieve stated goals, these objectives are almost always limited to maintaining a no net loss of wetlands and do not extend to specific measures, such as water quality levels or acreage of protected habitat. More specific objectives to actualize broad statements are needed to strengthen the ability of local plans to manage ecological systems.

While the basic intent to coordinate beyond jurisdictional and organizational boundaries is strong, the plans lack the building blocks of coordination. More specific collaborative techniques and detailed descriptions would perhaps foster more directed coordination necessary to protect transboundary resources. Nevertheless, the Inter-organizational Coordination and Capabilities plan component is particularly strong compared to others in the ecosystem planning protocol. These findings may be caused by recognition that managing coastal resources requires collaboration that does not necessarily adhere to human defined boundaries together with the State's requirements of an intergovernmental coordination plan component.

Overall, the policies, tools, and strategies plan component focuses primarily on a narrow set of traditional regulatory land use tools. A greater reliance on more innovative practices, particularly those based on incentives rather than strict regulation, would allow communities to expand their growth management toolbox, increase the quality of their plans, and more effectively manage ecological systems. Finally, the Implementation plan component falls short when it comes to making the policies "stick." One of the most frequently vocalized criticisms of plans in Florida is that they are not fully implemented after adoption.

Based on the above analysis, when held up against a model, local plans in Florida do not do particularly well at achieving the principles of ecosystem management. Aside from a few standout jurisdictions, such as Pinellas County, plans in general receive a poor grade which is an important warning sign when it comes to protecting the State's critical natural resources over the long term. Based on the empirical evidence, third and fourth generation local plan in Florida will need to strengthen and fact base, goals, and objectives components and in general better take into consideration the needs of entire ecological systems.

CASE 2: Identifying Policy Gaps Using Geographic Information Systems

This case study evaluates the collective capabilities of local jurisdictions to manage large transboundary ecological systems from a spatial point of view. Specifically, it uses Geographic Information Systems (GIS) to map, measure, and analyze the existing mosaic of management based on policies in comprehensive plans across selected ecosystems in the southern portion of Florida. Using GIS provides a spatial perspective on how well local plans are managing ecological systems in a way that the previous case is unable to accomplish. By examining the plans for multiple jurisdictions within large ecosystems, this case provides answers to the following research questions: 1) what is the degree of spatial coverage of ecosystem strategies and policies in southern Florida; 2) what is the existing spatial pattern of select indicators within specific ecological systems; 3) how well are multiple local jurisdiction collectively managing larger ecosystems; and 4) is the strength of ecosystem management capabilities randomly distributed across the study area, or clustered within particular ecological units? Visual and statistical results indicate significant gaps in the management framework of southern Florida that, if filled, could achieve a greater level of consistency and more complete coverage of ecosystem management policies.

The Sample

Sixteen adjacent ecosystem management areas (EMAs) defined primarily by watershed boundaries were selected for analysis in the southern portion of Florida (Figure 5.2). These ecosystems stretch from the west coast near Tampa Bay to the heavily developed southeast coast of the state, representing a wide variety of biophysical regions and institutional/political settings. Of particular importance is the Everglades system south of Lake Okeechobee, which is considered one of the most biologically diverse and valued natural system in the US while at the same time is being negatively impacted by increasing urban development. A sample of local jurisdictions was selected from among those jurisdictions containing land area within one of the sixteen EMAs. All counties intersecting the EMAs, plus the 15 largest cities in land area (since the goal is to achieve the greatest level of spatial coverage, cities were selected based on area rather than by population) were selected to form a sample of 45 adjacent local jurisdictions (Figure 5.3). On average, there are 6.35 jurisdictions within an EMA.

Scoring the Plans

The most recent comprehensive plans as of 2004 for these counties and cities were evaluated against the ecosystem planning protocol (presented in Chapter 4) to determine their collective ability to manage EMAs.[1]

---

1 Plans were evaluated by two trained coders working independently of each other. An "inter-coder reliability score" was computed equal to the number of coder agreements for

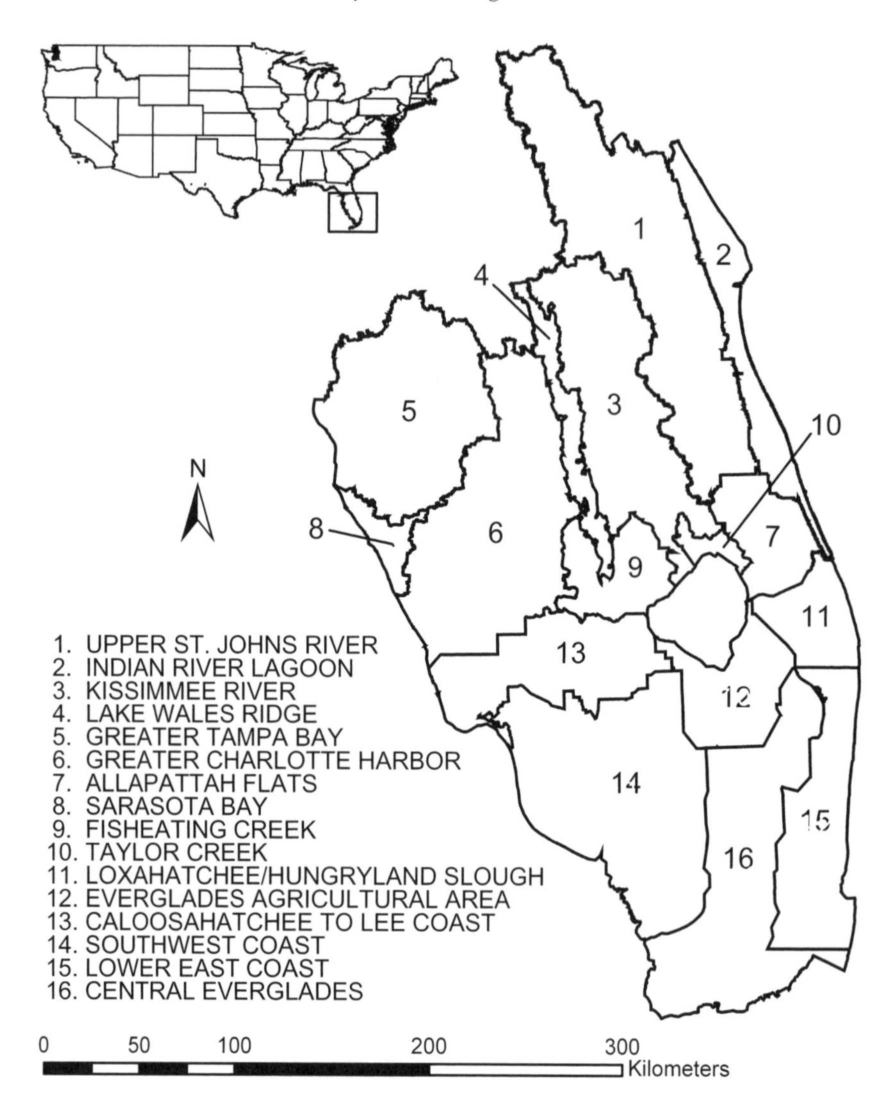

Figure 5.2 Ecosystem management areas
Source: Brody et al., 2003: 666.

---

indicators divided by the total number of indicators. We calculated a score of 97 percent. The literature suggests that an inter-coder reliability score in the range of 90 percent is generally considered acceptable (Miles and Huberman, 1984).

Figure 5.3 Local jurisdictions within the boundaries of EMAs
Source: Brody et al., 2003: 667.

The comprehensive plan for each jurisdiction located within a selected EMA was evaluated against 58 ecosystem management indicators contained in the coding protocol. If a policy or coordination strategy in a plan was mandatory (using the words "must," "shall," or "will"), it was coded as a 1. If the indicator was either suggested, faintly present, or not present at all, it was coded as a 0. With this method, the spatial coverage for each indicator could be mapped and measured across the sample of EMAs, spatially identifying gaps in the regional framework of management across southern Florida. As in the previous case study, it is important to note that plans are guides for future ecosystem management as opposed to how these strategies are implemented after the plans are adopted. However, one can assume that higher scoring plans have a greater likelihood of being implemented because only mandatory policies and programs were coded. Because local comprehensive plans in Florida are legally binding instruments, it is logical that their contents will be put into place. Jurisdictions in Florida have been sued by the state when their plans were found to be in noncompliance.

The percentage of spatial coverage for each indicator within each EMA was measured in two stages. First, we computed the percentage (P_{ij}) of the areas in the ith EMA that was occupied by the jth jurisdiction. Second, we used this proportion to weight that jurisdiction's contribution to the EMA's score on that indicator (I_{ijk}). That is,

$$P_{ik} = \sum_{j} P_{ij} I_{ijk}. \qquad (1)$$

Next, we computed a total ecosystem plan quality score (ΣEPQ) for each EMA in a similar manner. Specifically, we summed across all of the indicators for a given jurisdiction's plan and normalized this score by dividing by the total number of indicators and multiplying by 10 to place the variable on a 0-10 scale (as done in Case 1). Then, we weighted each jurisdiction's ΣEPQ score by the proportion (Pij) of the EMA's area occupied by that jurisdiction and summed across all jurisdictions within that EMA. That is,

$$EPQ_i = \left(\frac{10}{58}\right) \sum_{k} P_{ijk} \qquad (2)$$

Once plan quality scores for each EMA was calculated and entered into a GIS database, we conducted the following analysis. First, we used a visual and statistical gap analysis of ΣEPQ scores to reveal the mosaic of management across ecosystems in Florida. Specifically, we mapped 58 indicators and total ecosystem management scores (ΣEPQ) to facilitate an examination of both gaps in protection and spatial consistency of policies at a broad scale. Second, we investigated the spatial association of total ecosystem planning scores across the study area to determine if there is a clustered pattern of strong or weak ecosystem management capabilities.[2]

2 We used a Joint Count Statistic and Global Moran's I to form an overall picture of the degree of spatial dependency across the study area. We then employed a local indicator of spatial autocorrelation (LISA) (Anselin, 1995) to identify and map specific areas of spatial clustering. LISAs detect significant spatial clustering around individual locations and pinpoint areas that contribute most to an overall pattern of spatial dependence. This technique offered a finer focus to uncover important features or characteristics in explaining ecosystem management capabilities at the local level.

Finding the Gaps

Overall Pattern of Spatial Coverage

Tables 5.7 and 5.8 report the percentage of spatial coverage of detailed or mandatory ecosystem indicators for each EMA and across the entire study area. We assume an indicator with a total spatial coverage of below 20 percent represents an insufficient degree of consistency between multiple jurisdictions and a potential gap in the ecosystem management framework across southern Florida. With respect to the *Inter-organizational Coordination and Capabilities Plan Component* (Table 5.7), financial commitment to ensure implementation of collaborative strategies is covered by just over 1 percent of the study area. Collaboration between science and policy organizations (an essential aspect of managing ecological systems) receives less than 6 percent of total coverage, concentrated primarily in south Florida EMAs. In addition, intergovernmental agreements to protect the integrity of ecosystems cover approximately 13 percent of the study area. Some of the highest levels of coverage for this indicator occur in South Florida/Kissimmee River (50.72 percent) and Greater Tampa Bay (36.73 percent) EMAs where formal regional agreements, such as a National Estuary Program (NEP) are already in place. Finally, participation in ecosystem-based initiatives is specified in the plans of approximately 16 percent of the sample of local jurisdictions. Many large ecosystem programs, such as NEPs and EMA planning initiatives rely on the participation of cities and counties for successful implementation. The greatest intent to participate from a spatial perspective comes from the South Florida EMAs most directly associated with the Everglades region. In contrast to the weaknesses in the collaborative management framework across the study area, nearly 92 percent of the sample specifies coordination with Water Management Districts and approximately 80 percent of all jurisdictions integrate other regional plans and policies into their local comprehensive plans.

Within the *Policies, Tools and Strategies* plan component (Table 5.8), two important policies for protecting ecosystem components that are absent from all plans in the sample are phasing of development to reduce wildlife disturbance and the designation of urban growth boundaries that do not include critical habitat (Deurksen et al., 1997). Policies associated with public funding strategies, such as controlling investment for public projects and capital improvements programming to protect ecosystem components also receive weak mandatory coverage across the study area. Interestingly, EMAs to the north and west of Lake Okeechobee contain the highest concentrations of public funding policies.

Finally, mandatory tax-based policies encouraging development patterns that protect critical habitats and ecosystem processes cover less than 20 percent of the study area. Policies involving preferential tax treatments to protect critical habitats and the designation of special taxing districts to raise funds for land acquisition are found almost entirely in the South Florida Loxahatchee/Hungry Slough and Everglades Agricultural EMAs. The highest level of spatial coverage and consistency across EMAs is associated with more traditional land use policies, such as use restrictions in and around critical habitats, protection of endangered and threatened species, and the protection of native vegetation.

Table 5.7 Ecosystem management indicators (P_{ik}) for inter-organizational coordination and capabilities

Policies	Upper St. Johns River	Indian River Lagoon	S. Florida Kissimmee River	Lake Wales Ridge	Greater Tampa Bay	Greater Charlotte Harbor	S. Florida Allapattah Flats	Sarasota Bay	S. Florida Fisheating Creek
Commitment of financial resources	9.27	0.00	0.91	5.35	8.48	0.00	0.00	0.00	0.00
Link betw. Science and policy	0.00	0.34	0.00	0.00	8.48	0.00	30.92	0.00	0.00
Position in ecosystem ID'd	13.97	0.00	14.27	0.00	14.06	14.54	0.00	0.00	1.01
IGAs specified	38.60	18.01	50.72	2.83	36.73	8.01	0.29	25.59	0.00
Participation in Ecosyst. initiatives	32.38	42.33	0.00	0.00	8.48	0.00	0.00	0.00	0.00
Joint database production	24.68	9.34	18.00	57.52	18.21	36.50	30.92	0.00	1.01
Intergovernmental bodies	31.80	19.68	13.20	0.00	36.37	21.86	95.84	15.93	1.01
Coord. with private sector	35.25	21.42	55.51	8.17	19.82	46.20	10.23	67.07	1.01
Other coord.	54.67	64.10	0.00	0.00	30.94	26.57	39.07	15.93	0.00
Coord. within jurisdiction	17.73	21.77	18.00	57.52	35.85	48.54	39.07	15.93	0.00
Coord. with federal	37.41	28.35	30.36	45.01	64.38	17.40	65.21	9.66	99.76
Conflict management processes	25.35	18.35	18.00	57.52	41.79	30.04	31.21	15.93	0.00
Information sharing	49.57	12.76	32.27	57.52	41.03	30.04	38.78	15.93	0.00
Coord. with other organizations	66.40	16.27	63.86	47.83	36.87	9.98	95.84	15.93	54.29
Other organizations ID'd	44.86	28.69	52.58	2.83	30.94	21.96	96.13	25.59	45.46
Coord. with state	47.85	28.69	30.36	45.01	65.08	49.76	96.13	76.73	99.76
Coord. with adjacent jurisdictions	69.19	42.33	100.00	94.65	53.13	81.85	9.94	92.66	98.76
Integration with other plans	67.13	70.68	51.33	97.18	98.37	59.73	75.15	92.66	99.76
Coord. with WMDs	89.30	71.02	100.00	100.00	75.56	90.92	100.00	76.73	99.77

Table 5.7 continued

Policies	S. Florida Taylor Creek	S. Florida Loxahatchee/ Hungryland Slough	S. Florida Everglades Agricultural	Caloosahatchee to Lee Coast	Southwest Coast	S. Florida Lower East Coast	S. Florida Central Everglades	Avg Policy Coverage
Commitment of financial resources	0.00	0.00	0.00	0.00	0.00	1.96	0.00	1.62
Link betw. Science and policy	14.31	32.75	0.38	0.00	0.00	0.00	0.00	5.45
Position in ecosystem ID'd	0.00	0.00	0.00	18.76	1.41	36.74	45.14	9.99
IGAs specified	0.00	0.00	0.00	6.65	13.28	0.00	12.85	13.35
Participation in Ecosyst. initiatives	0.00	64.82	84.54	0.00	0.00	18.79	10.06	16.34
Joint database production	14.31	40.08	0.38	12.47	55.75	2.16	0.00	20.08
Intergovernmental bodies	19.79	97.56	84.92	18.76	0.00	18.79	10.06	30.35
Coord. with private sector	79.71	72.14	84.54	12.11	0.00	18.99	10.06	33.89
Other coord.	14.31	97.56	84.92	7.01	69.03	23.03	22.90	34.38
Coord. within jurisdiction	14.31	100.00	85.16	0.00	3.21	83.16	80.59	39.11
Coord. with federal	5.47	7.33	2.69	66.58	25.13	66.64	83.38	40.92
Conflict management processes	14.31	97.56	97.19	60.87	40.29	57.81	68.04	42.14
Information sharing	14.31	100.00	84.92	29.39	79.08	55.73	68.04	44.64
Coord. with other organizations	19.79	97.56	97.19	60.87	27.01	61.76	55.19	51.67
Other organizations ID'd	19.79	100.00	99.64	63.10	87.40	62.24	68.04	53.08
Coord. with state	19.79	100.00	87.37	66.58	23.32	62.24	68.04	60.42
Coord. with adjacent jurisdictions	79.71	64.82	99.26	80.03	96.04	57.81	68.04	74.27
Integration with other plans	85.18	72.14	99.26	63.10	87.40	61.96	68.04	78.07
Coord. with WMDs	99.49	100.00	99.88	92.14	97.85	89.70	93.44	92.54

Source: Adapted from Brody et al., 2003: 668.

Table 5.8 Ecosystem management indicators (P_{ik}) for policies, tools, and strategies

Policies	Upper St. Johns River	Indian River Lagoon	S. Florida Kissimmee River	Lake Wales Ridge	Greater Tampa Bay	Greater Charlotte Harbor	S. Florida Allapattah Flats	Sarasota Bay	S. Florida Fisheating Creek
Urban growth boundaries	0.00	0.00	0.00	0.00	0.00	0.00	0.00	0.00	0.00
Phasing development	0.00	0.00	0.00	0.00	0.00	0.00	0.00	0.00	0.00
Control public investments	26.24	33.34	0.91	5.35	14.42	0.00	0.00	0.00	0.00
Subdivision standards	0.00	0.34	0.00	0.00	0.00	0.00	30.92	0.00	0.00
Other incentives	22.97	18.01	0.91	5.35	64.73	33.84	0.29	0.00	1.03
Targeted growth areas	21.13	8.99	0.00	0.00	16.88	26.53	0.00	15.93	0.02
Capital improvements	9.75	3.41	0.91	5.35	8.48	14.60	0.00	0.00	1.01
Fencing controls	0.00	0.00	0.00	0.00	0.00	0.00	0.00	0.00	0.00
Preferential tax treatments	0.00	0.34	0.00	0.00	0.00	0.00	30.92	0.00	0.00
Special taxing districts	0.00	0.34	0.00	0.00	8.48	0.00	30.92	0.00	0.00
Mitigation banking	14.99	0.00	0.91	5.35	19.49	14.60	0.00	0.00	1.01
Network protected areas	32.66	27.00	18.00	57.52	18.21	21.97	0.29	0.00	0.00
Impact fees	15.41	16.26	0.00	0.00	0.37	0.69	87.98	0.00	0.00
Structural solutions	31.96	33.34	0.91	5.35	8.48	14.54	0.00	0.00	1.01
Density bonuses	43.97	8.99	27.42	45.01	47.79	2.70	0.00	0.00	54.29
Performance Zoning	56.37	42.67	38.43	8.17	18.34	37.94	30.92	83.00	0.02
Other land acquisition	39.95	27.35	1.07	0.00	53.37	19.42	39.07	0.00	0.00
Public access controls	17.45	44.01	0.00	0.00	42.63	32.59	87.98	92.66	1.01
Fee simple purchase	42.68	27.34	26.51	39.66	42.63	22.34	31.21	83.00	54.29

Table 5.8 continued

Policies	S. Florida Taylor Creek	S. Florida Loxahatchee/ Hungrylandslough	S. Florida Everglades Agricultural	Caloosahatchee To Lee Coast	Southwest Coast	S. Florida Lower East Coast	S. Florida Central Everglades	Avg Policy Coverage
Urban growth boundaries	0.00	0.00	0.00	0.00	0.00	0.00	0.00	0.00
Phasing development	0.00	0.00	0.00	0.00	0.00	0.00	0.00	0.00
Control public investments	0.00	0.00	0.00	0.00	0.00	0.00	0.00	5.02
Subdivision standards	14.31	32.75	0.38	0.00	13.28	0.00	12.85	6.55
Other incentives	0.00	0.00	0.00	12.11	0.00	0.00	0.00	9.95
Targeted growth areas	0.00	0.00	12.27	25.19	30.25	0.00	12.85	10.63
Capital improvements	0.00	0.00	0.00	41.14	10.04	36.74	45.14	11.04
Fencing controls	0.00	64.82	84.54	0.00	13.28	18.79	22.90	12.77
Preferential tax treatments	14.31	97.56	84.92	0.00	0.00	18.79	10.06	16.06
Special taxing districts	14.31	97.56	84.92	0.00	0.00	18.79	10.06	16.59
Mitigation banking	0.00	64.82	84.77	41.14	10.45	46.24	35.46	21.20
Network protected areas	0.00	64.82	84.54	0.00	0.00	18.79	10.06	22.12
Impact fees	19.79	97.56	84.92	0.00	13.28	18.79	22.90	23.62
Structural solutions	0.00	64.82	97.05	37.30	18.77	44.25	35.46	24.58
Density bonuses	0.00	64.82	84.54	0.00	0.00	18.79	10.06	25.52
Performance Zoning	14.31	32.75	12.66	54.23	25.60	0.00	0.00	28.46
Other land acquisition	14.31	97.56	84.92	35.68	21.92	18.79	22.90	29.77
Public access controls	19.79	32.75	0.38	48.15	79.08	38.71	57.98	37.20
Fee simple purchase	14.31	104.89	84.92	0.36	55.75	18.99	10.06	41.18

Source: Adapted from Brody et al., 2003: 670.

Spatial Analysis of Specific Ecosystem Management Indicators

While examining the spatial coverage of ecosystem management indicators across the entire study area provides a general idea of policy gaps or weaknesses, it does not indicate the degree of coordination within the EMAs themselves. We therefore selected four indicators for a more detailed examination of their spatial distribution. Specific coordination capabilities and policies in the sample of comprehensive plans were mapped and analyzed according to their respective EMAs. Through these four examples, we could more precisely identify deficiencies in the spatial coverage of important ecosystem management capabilities across multiple local jurisdictions. The examples also demonstrate the effectiveness of using GIS techniques to assess ecosystem management capabilities at the local level and identify area specific gaps in management at the ecosystem level.

Actions to Protect Natural Resources Crossing Into Other Jurisdictions: Mapping jurisdictions that have policies to coordinate with their neighbors and neighbor's neighbors within each EMA reveals gaps in the management framework at the watershed level. For example, in the Great Tampa Bay Area EMA, Pinellas, St. Petersburg, Pasco, and Manatee all have mandatory collaborative policies to help protect ecosystem components in neighboring jurisdictions, while Hillsborough and Polk counties do not have such a policy in their comprehensive plan (Figure 5.4). Similarly, within the Southwest Coast EMA, Lee and Monroe Counties have policies to protect resources crossing into adjacent jurisdictions, but such policies are absent in the plans of Hendry, Collier, and Broward Counties. In both instances, mapping this indicator by EMA shows where local and regional planners need to set local policies to attain a more complete spatial coverage of riverine and coastal ecological communities.

Sharing Information with Other Organizations: Table 5.7 shows that information sharing covers almost 50 percent of the study area, but mapping this indicator shows important spatial gaps within several EMAs (Figure 5.5). For example, jurisdictions in the northern portion of the St. Johns River EMA (a watercourse flowing into Lake Okeechobee), contain detailed commitments to share information pertaining to the management of ecological systems. However, the ecosystem management indicator is missing from the plans of jurisdictions to the south. Furthermore, while the comprehensive plans for Dade and Palm Beach Counties (the anchor jurisdictions for South Florida Lower East Coast and Central Everglades EMAs) contain information sharing strategies, Broward County's plan does not.

The absence of detailed information sharing policies for Broward thus acts as significant policy gap and potential barrier for the effective flow of information from one jurisdiction or government organization to the next. If Broward County adopts information sharing policies in the next update of their plan, the exchange of vital ecological data, collaboration between jurisdictions, and the management of the ecosystems could be significantly enhanced.

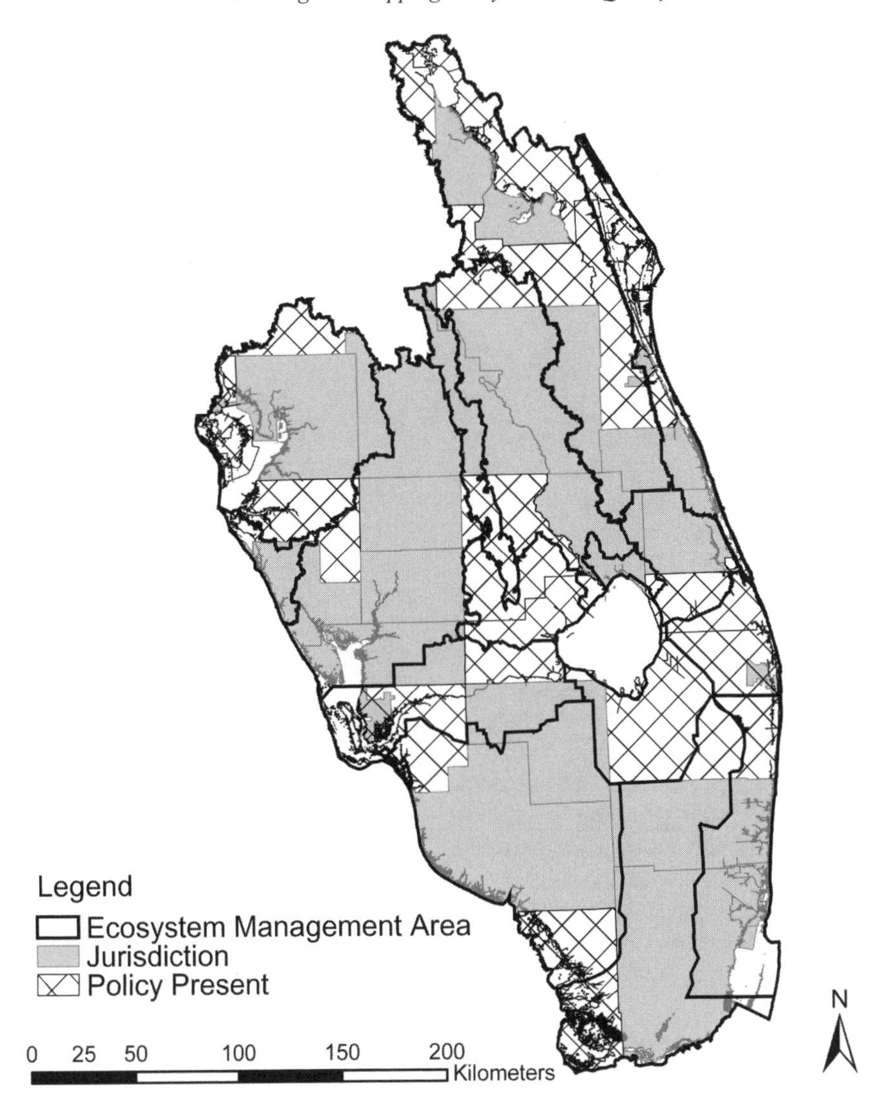

Figure 5.4 Distribution of policies for actions to protect natural resources crossing into adjacent jurisdictions

Source: Brody et al., 2003: 673.

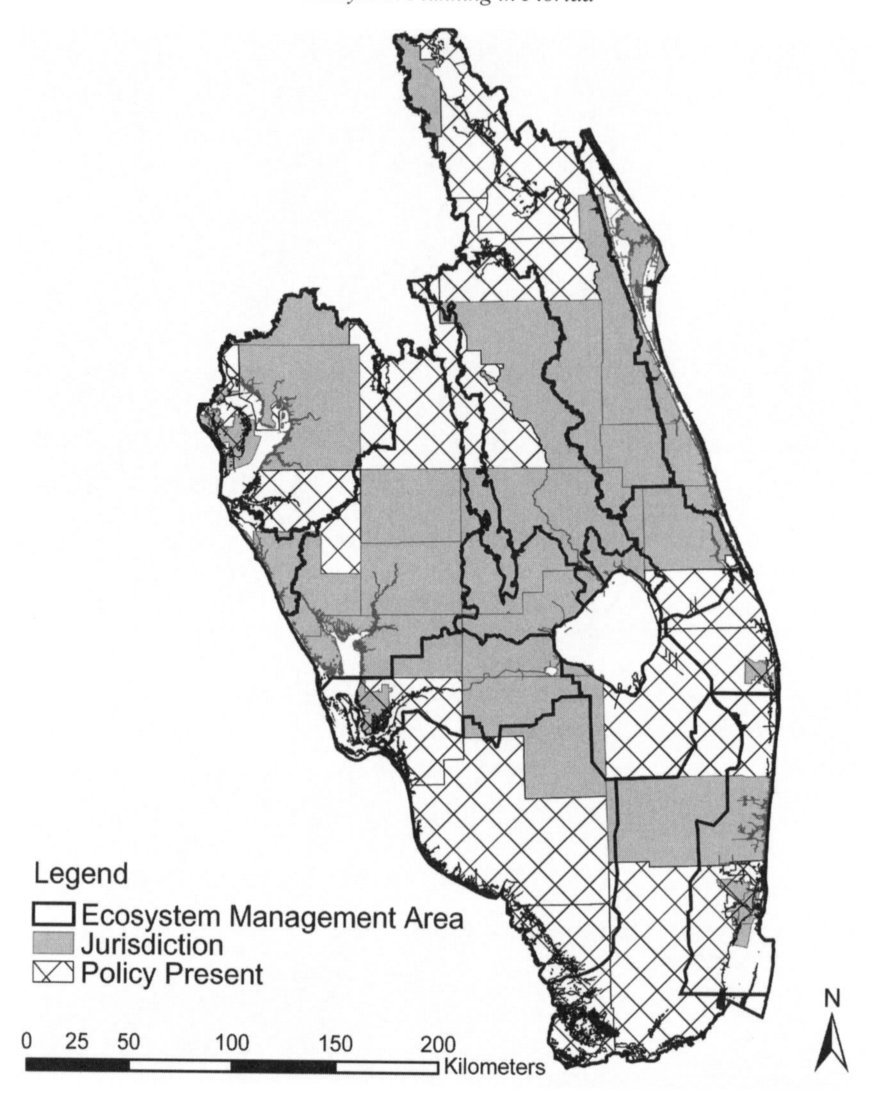

Figure 5.5 Distribution of strategies for information sharing
Source: Brody et al., 2003: 674.

Establishment of Wildlife Corridors: One of the most important concepts originating from the landscape ecology and conservation biology literatures is that habitats do not stand alone, but are connected by the movement of species, water, and natural materials. A spatial analysis of environmental policies illuminates gaps in county and municipal agencies that support the maintenance of corridors. Mandatory policies for the establishment or protection of wildlife corridors are found in the plans of approximately 50 percent of the study area. Notable policy gaps exist in the central portions of the Upper St. Johns River, South Florida Lower East Coast and

Central Everglades EMAs (Figure 5.6). One of the largest gaps and lack of spatial coverage for corridors, however, occurs in the Southwest Coastal EMA. Lee County to the northwest portion of this ecosystem appears to be the only jurisdiction with significant area within the EMA containing mandatory wildlife corridor policies in its plan. This policy gap in and around Collier County potentially leaves panther (*Puma concolor coryi*), black bear (*Ursus americanus floridanus*), and bobcat (*Lynx rufus floridanus*) unprotected by local comprehensive planning frameworks.

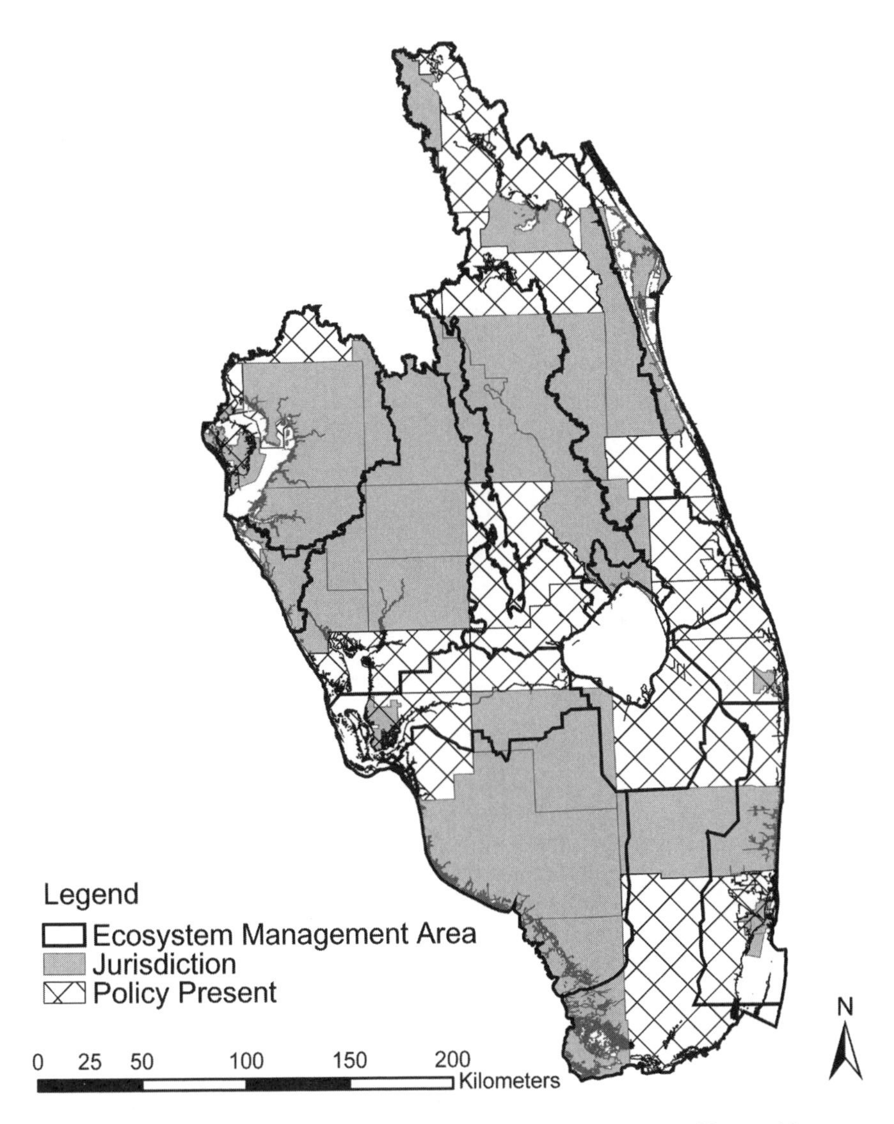

Figure 5.6 Distribution of policies to maintain or create wildlife corridors
Source: Brody et al., 2003: 675.

Controls on Exotic and Invasive Species: Exotic or non-native species can greatly alter the composition, structure, and function of ecological systems if left unchecked. Due to the seriousness and general recognition of this environmental problem, local policy controls on exotic and invasive species are covered by almost 70 percent of the study area (see Table 5.8). Jurisdictions in the South Florida EMAs south of Lake Okeechobee, where the introduction of exotic species is greater than any other drainage area in the State, appear to pay particular attention to the issue in their comprehensive plans.

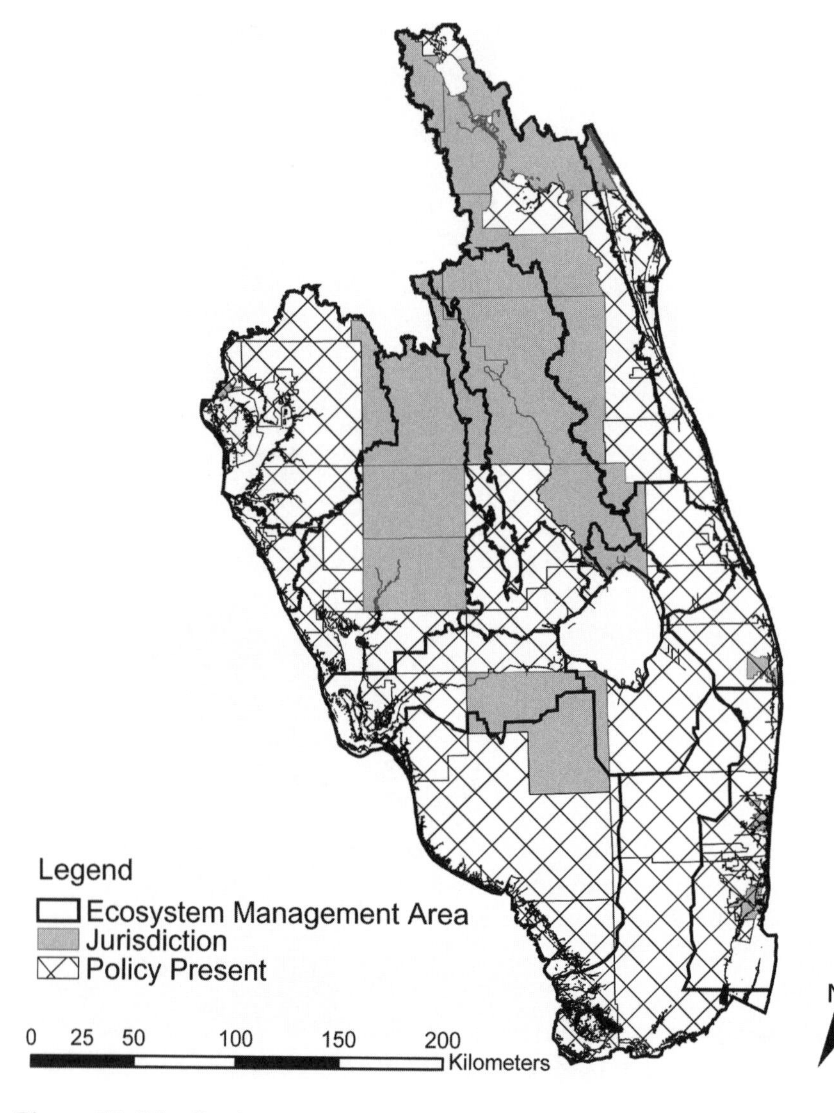

Figure 5.7 Distribution of policies to control or remove exotic species
Source: Brody et al., 2003: 676.

Despite wide coverage of this ecosystem management indicator, important policy gaps do exist (Figure 5.7). For example, the absence of exotic controls in the Hendry County comprehensive plan results in significant gaps in the Calooosahatchee to Lee Coast and the Southwest Coast EMAs. Polk, Hardee, and De Soto Counties in the western part of the Greater Charlotte Harbor EMA are also missing this important ecosystem planning policy in their local management frameworks.

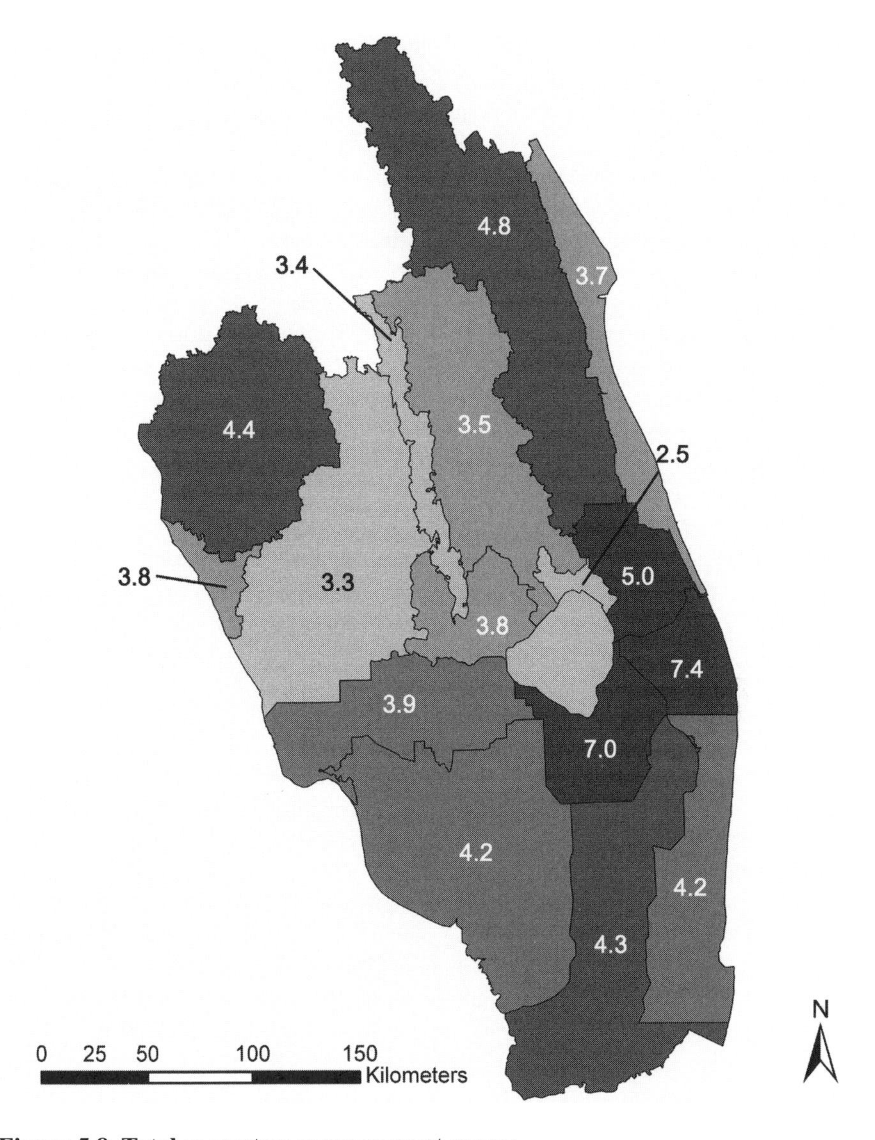

Figure 5.8 Total ecosystem management scores
Source: Brody et al., 2003: 677.

Evaluation of Total Ecosystem Planning Scores

The ΣEPQ score computed from the area-weighted sum of all indicators allowed us to evaluate the overall management capabilities of multiple jurisdictions for each EMA. This phase of analysis serves as a global assessment of the relative strength of the potential for ecosystem management. In general, the highest total ecosystem planning scores occur for coastal EMAs containing areas of valued critical natural resources that are high priorities for protection (Figure 5.8). For example, EMAs covering the well-known Everglades region south of Lake Okeechobee all receive above average scores. The Greater Tampa Bay EMA, a similarly well-known ecosystem with multiple environmental initiatives including a NEP, also scored comparatively high.

The highest scores, however, appear in the South Florida Everglades Agricultural and Loxahatchee /Hungryland Slough EMAs along the southeast coast of the State. The ecosystem planning capabilities for these adjacent EMAs are well above average, primarily due to the strength of the Palm Beach County plan, which is one of the highest scoring jurisdictions among the sample of comprehensive plans. In contrast, inland EMAs with lower levels of biodiversity and lesser-known natural value receive lower scores. The Upper St. John River and South Florida Appattah Flats EMAs are two notable exceptions.

Evaluation of Spatial Distribution of Local Ecosystem Planning Scores

While the measure and comparison of ΣEPQ scores for each EMA enabled us to make overall assessments of ecosystem management capabilities, it could not provide detail on areas of high quality management within the EMAs themselves. In the final phase of the descriptive analysis, we used measures of spatial autocorrelation to assess the geographic distribution of ΣEPQ scores. This approach allowed us to determine if there is significant spatial association of plan scores for adjacent jurisdictions across the study area and identify clusters of high or low scores within specific EMAs.

We conducted three tests for spatial autocorrelation: a Joint Count Statistic, a Global Moran's I and a Local Moran's I (Table 5.9). All three tests indicate statistically significant spatial autocorrelation among local jurisdictions ($p<.05$) as well as a tendency for high ecosystem planning scores to cluster geographically within EMAs. Both the Joint Count and Global Moran's I statistics show a significantly nonrandom or clustered pattern of scores across the entire study area. Mapping the most statistically significant Local Moran's I scores ($p<.05$) enabled us to identify the location of clusters or hot spots of adjacent jurisdictions with high ΣEPQ scores.

These jurisdictions serve as spatial hot spots of significantly greater capabilities to manage ecological systems over the long term. For example, a particularly strong clustered pattern of high ecosystem planning scores occurs in the Greater Tampa Bay EMA. The comprehensive plan for Pinellas County in the western part of this watershed is distinguished for its emphasis on ecosystem management and the protection of critical habitats. This plan received the highest total score in the sample of jurisdictions.

Table 5.9 Tests for spatial autocorrelation for ecosystem planning scores

Statistical Test	Expected	Observed	Z-Value	Significance
Joint-count*				
1-1	36.870	71.500	0.567	0.571
0-0	90.646	101.000	0.068	0.945
1-0	117.984	73.000	-5.406	0.000
Moran's I	-0.010	0.155	2.269	0.023
LISA	-0.009	0.470	2.170	0.030
	-0.009	1.240	3.850	0.000
	-0.009	1.390	3.200	0.001
	-0.009	1.280	2.290	0.022
	-0.009	1.110	2.290	0.022
	-0.009	1.110	2.290	0.022
	-0.009	1.010	4.130	0.000
	-0.009	1.300	4.690	0.000
	-0.009	1.800	3.210	0.001
	-0.009	0.780	1.990	0.046

Source: Adapted from Brody et al., 2003: 678.

Note: * = Total ecosystem planning scores were converted into a dichotomous variable (above or below average) to obtain a measure of spatial autocorrelation using joint count statistics.

The same situation occurs on the east coast of the state with the case of Martin County. Martin is the third highest scoring plan in the sample and is adjacent to Palm Beach, which received the second highest score. Together, these jurisdictions appear to be partly responsible for the high $\sum$EPQ scores for the Everglades Agricultural Area and Loxahatchee /Hungryland Slough EMAs.

Filling the Gaps

By graphically unfolding the spatial pattern of local ecosystem policies across southern Florida, we provide guidance to policy makers and planners interested in strengthening the ability of communities to effectively manage the larger natural system within which they are situated. In general, we find that incentive and financial based strategies, such as preferential tax treatments and special funding programs are lacking in coverage compared to more tradition land use regulatory tools (see also Brody, 2003). It is not surprising that financial incentives are found almost entirely in the South Florida Loxahatchee/Hungry Slough and Everglades Agricultural EMAs which is where some of the most intense development and private investment in the State occur. In contrast, traditional regulatory land use policies have the greatest level of spatial coverage. These types of policies are readily incorporated into local plans in part because they receive less opposition than other more complicated or untested planning tools. They are also easy to implement and enforce.

In terms of the spatial distribution of indicators within specific EMAs, notable gaps persist in the Southwest Coast, Southeast Coast, and Central Everglades EMAs, particularly for collaboration with neighboring jurisdictions and wildlife corridors. For example, the Tampa Bay watershed is a well-defined ecological system managed by a National Estuary Program (NEP) plan. However, it is equally important that local jurisdictions within this EMA not only participate in the ecosystem plan, but also incorporate specific policies into their legally binding land use instruments that require government to realize that natural systems do not adhere to a single political or administrative entity. Without the regulatory commitment of local jurisdictions within the Tampa Bay watershed, regional environmental programs such as a NEP will be largely ineffective in attaining their goals.

With respect to the establishment or protection of corridors for panthers, black bears and other highly mobile species, the Southwest Coast EMA contains a potentially significant gap in and around Collier County. Efforts have in fact been made by state and federal organizations to provide natural corridors in Florida. For example, the State Department of Transportation and US Fish and Wildlife Service have constructed faunal underpasses along Interstate Highway 75 linking Fort Lauderdale and Naples to allow panther and other mobile species to move unencumbered along natural corridors (Smith, 1993; Foster and Humphrey, 1995).

While there have been some exceptional initiatives to maintain wildlife corridors, our analysis shows that in southwest Florida, wildlife corridor policies and related projects have not been adequately incorporated into local land use decisions where they may have the most significant impact. Figure 5.9 illustrates that the lack of local wildlife corridors in Collier County may have a significant impact on the viability of panther populations in Florida. The majority of panther habitat in the State occurs where major gaps exist for wildlife corridor policies. The same area is where the highest concentration of panther kills (primarily along roadways) has been reported. Filling the regulatory gap for wildlife corridors by updating the Collier County plan may reduce the number of panther deaths within this area.[3]

---

3 Subsequent to this assessment, the Collier County Growth Management Plan was amended (in late 2002 and 2004) to include mandatory policies for wildlife corridors.

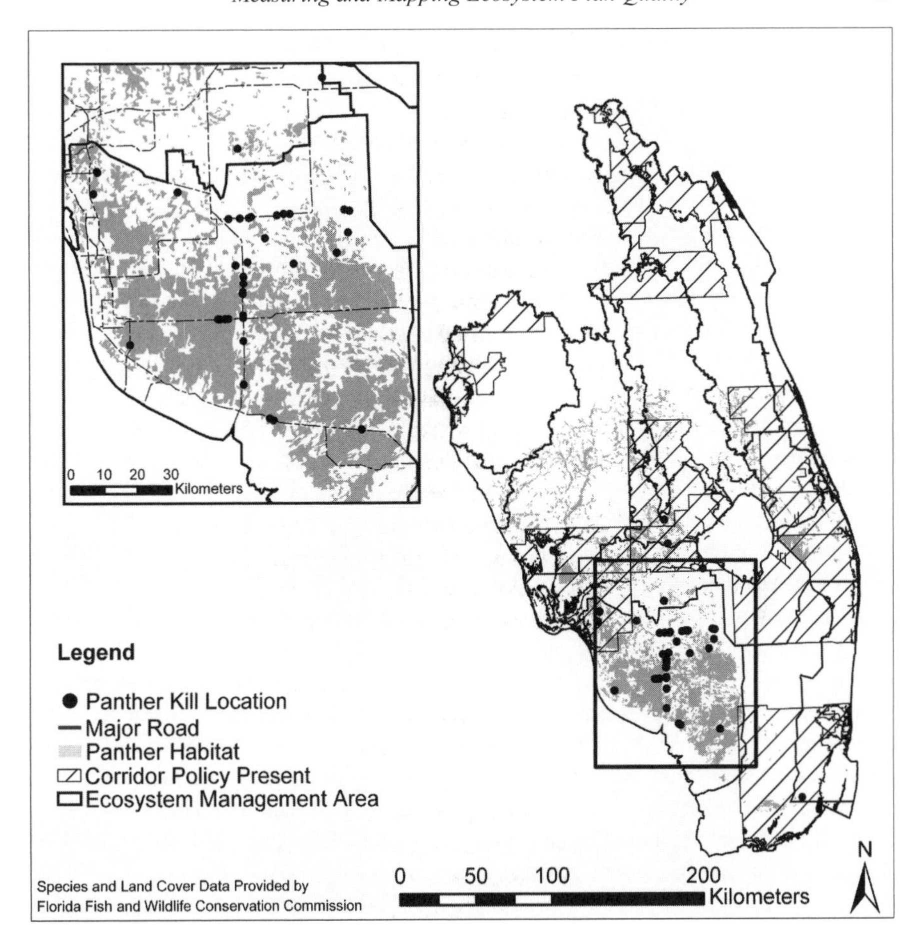

Figure 5.9 Florida panther habitat and panther kills
Source: Brody et al., 2003: 679.

The results of this case also demonstrate significant gaps in policies related to the removal of exotic species. For example, the Giant Rams-horn Snail (*Marisa cornuarietis*) is just one well-known exotic species observed in Hendry County. The snail is native to northern South America and was most likely released by aquarium hobbyists. Studies show that this species retards the growth of aquatic plants by feeding on the roots of the plants (Fuller and Benson, 1999). The lack of controls in Hendry County could be influential in the Snail's establishment in neighboring Palm Beach and Broward Counties. In another example, the Cuban Treefrog (*Osteopilus septentrionalis*) has been established in Polk County for at least ten years. This species preys upon smaller native treefrogs and may reduce their populations via competition and predation (Asthon and Asthon, 1988). Brazilian Pepper (*Schinus terebinthifolius*) is another exotic species established in Polk and De Soto counties that was originally imported from Brazil as an ornamental in the 1840s. This plant species forms dense thickets of woody stems that can shade out and displace native vegetation. It also produces certain allelopathic agents, which appear to suppress other plant's growth (Mahendra et al., 1995). Brazilian Pepper is now estimated to occupy over 283,400 ha in central and south Florida alone (Ferriter, 1997). In general, it is important for adjacent local jurisdictions to implement consistent exotic control strategies since nonnative species can spread rapidly across large areas.

By measuring and mapping ∑EPQ scores, results show that if planners wish to improve the lower scoring ecosystems in southern Florida, they should focus ecosystem management efforts on less developed, inland-dominated EMAs. This recommendation would entail a more proactive approach to local planning, where ecosystem protection strategies are established before significant development takes place, as opposed to a traditional reactionary policymaking stance.

Finally, the discovery that high quality plans tend to occur in neighboring jurisdictions, and that there are literally hot spots of strong ecosystem management capabilities has potential policy implications. Upon further review of spatial autocorrelation and comprehensive plans within the Greater Tampa Bay EMA, we observed that cities and counties adjacent to Pinellas seem to borrow from its plan environmental data, descriptions of programs, and wording of specific policies. This trend suggests that an extremely high scoring comprehensive plan may influence the quality of plans in neighboring jurisdictions through communication, collaboration, and information sharing. This phenomenon is most noticeable for cities located within a specific county. Another potential factor that seems to contribute to a spatial clustering of high scoring plans is that the Tampa Bay watershed has a history of collaborative environmental projects (most notably a NEP and associated plan), where there are previously established lines of communication, information sharing, and joint data production. The presence of a collaborative network may facilitate the dispersion of plan content, resulting in an intense hot spot of high scoring plans. Similarly, the strong strategies, programs, and policies in the Martin County plan appear to have diffused to neighboring jurisdictions and increased the quality of their plans. The result is a concentrated hot spot of high quality ecosystem-oriented local plans that contribute to the effective management of a large ecological unit.

In general, using GIS techniques to map, measure, and analyze the existing mosaic of management across ecological systems in southern Florida helps form a clearer picture of how local jurisdictions can join together to protect transboundary critical natural resources. Mapping specific strategies and policies provides a rapid assessment of ecosystem protection and provides a strategic tool with which to plan more effectively at the level of natural systems. By locating the spatial gaps (or policy weaknesses) within layers of planning policies and among groups of aggregated policies, the total resource management system can be better understood and strengthened through future plan updates.

Rather than casting a broad regulatory net, a spatially focused approach may be more precise, efficient, and cost effective. These techniques can be extremely useful for state and regional planners interested in managing large ecological units such as watersheds and assist local planners who understand protecting their own natural resource base requires focusing beyond their single jurisdiction.

PART 2
The Process: Factors Influencing Local Ecosystem Plan Quality

Part 1 of this book focused on the plan as a local level regulatory vehicle for capturing the principles of effective ecosystem management. We conceptually developed a model of what a local plan should contain if the overarching goal is to manage ecological systems that may extend beyond a single jurisdiction's boundaries. We then put this model to an empirical test by statistically and spatially evaluating it against actual Florida plans in two case studies.

Part 2 of this book focuses on the planning process. Given that a comprehensive plan in the US is not meant to be an isolated document written by technocrats behind closed doors, but instead influenced by specific local socioeconomic and biophysical characteristics and extensive community input, the process and context of plan-making must be considered. This Part concentrates on two major factors known to influence local ecosystem plan quality: 1) the degree of biodiversity within the planning area; and 2) the degree of stakeholder participation. Background on the major concepts is presented for each factor, as well as a brief explanation of how they may affect ecosystem plan quality. Then, several case studies are presented to help answer the research question: *what are the factors and processes influencing the quality of comprehensive plans with regard to ecological management?*

Specifically, Chapters 6 and 7 examine the role of the physical landscape and biological diversity ("biodiversity") in shaping the environmental content/quality of local plans. Two empirical cases are presented. The first measures the degree of biodiversity within a local jurisdiction and explains its impact on the quality of local plans. The second study uses GIS to analyze neighboring jurisdictions in the southern part of Florida and better understand how biodiversity and human disturbance of critical natural resources motivate planners to adopt high quality plans. Chapters 8 and 9 focus on the impact of public participation and stakeholder representation on local ecosystem plan quality. An in-depth case study is presented on the influence of various stakeholder groups, including resource-based industry on the quality of adopted local comprehensive plans.

Chapter 6

The Fragmented Landscape: Biological Diversity versus Human Disturbance

Landscape Structure and the Protection of Biodiversity

The protection of biological diversity is considered to be an overarching goal of ecosystem management (Grumbine, 1990; Grumbine, 1994; Slocombe, 1998; Noss and Scott, 1997; McCormick, 1999). Because species diversity is perceived as a fundamental component to maintaining viable ecosystems over the long term, the identification and protection of biodiversity is at the core of planning for ecosystem integrity (Vogt et al., 1997). Defined as "the full range of variety and variability within and among living organisms, and the ecological complexities in which they occur" (Peck, 1998; p. 189), biodiversity is often operationalized as species richness (the overlap of focal species). It is the intersection of key species that supports the overall function and processes of ecological systems. For this reason, planners have targeted biodiversity and its various components in their attempts to manage ecosystems.

Biodiversity occurs at various scales, including genetic, population, community and landscape (Noss, 1983; Noss and Cooperrider, 1994). Land managers usually focus on a scale that includes a series of interconnected or fragmented habitats, or regions that support sufficiently large areas of unaltered indigenous ecosystems. The motivation to protect biodiversity as a surrogate for maintaining ecosystem integrity is based on its value. First, it provides direct utilitarian values, such as food and medicine to human populations. Natural habitats have long provided local people with a means for survival in the form of meat and vegetables, firewood, medicinal plants, etc. Second, highly diverse natural systems, which support a wealth of species provide important "ecosystem services," such as maintaining hydrological cycles, regulating climate, storing and cycling essential nutrients, and absorbing and breaking down pollutants. Third, biodiversity provides sites for tourism, recreation, research, and other human activities (McNeely, 1992).

While it is beyond the scope of this book to detail all of the ecological principles underlying biodiversity, a few concepts are important for local planners to understand, particularly when it relates to making high quality plans. First, ecosystems are patchy in their abundance of life and location of biodiversity. Understanding how biodiversity is distributed across natural landscapes in the form of patches or habitat is essential to developing plans that seek to manage ecosystems. The literatures on landscape ecology and conservation biology provide the most insight into the components of ecosystems. The works of Foreman (1986; 1995a; 1995b) along with other authors are the most relevant studies for understanding the role of habitats in protecting biodiversity and the integrity of natural systems over the long term. A central theme emerging from this research is the concept of the *patch*. A patch is usually defined as a spatially separate instance of a given type of habitat. For example, a stand of aspens surrounded by conifers is a patch for some species of birds (see Duerksen et al., 1997 for a more detailed discussion of patches as they relate to land use planning). Patches are analyzed by size and location among other patches. A large patch is likely to have more habitats present and therefore contain a greater number of species than a small patch. Large patches support larger populations, which are more persistent in the face of human disturbance, such as suburban sprawl. Larger patches also contain more diversity of species than smaller ones. Overall, they protect aquifers and interconnected stream networks, sustain viable populations of species, provide core habitat and escape cover for large species, and permit near natural disturbance regimes (Dramstead, 1996).

The most important concept originating from the landscape ecology and conservation biology literatures is that habitats do not stand alone, but are connected by the movement of species, water, and natural materials. For example, a series of small patches can act as "stepping stones" for species movement. These interrelated patches provide the focus for management efforts aimed at ecosystem protection. Systems of interconnected habitats make-up the "landscape mosaic," which refers to the overall connectivity of natural systems (Soule, 1991; Peck, 1998). Connectivity is essential to maintain biodiversity and ecosystem integrity because it facilitates species movement and dispersal (Noss, 1991).

Second, while there is some debate over the issue, most ecosystem ecologists assert that *corridors* are an essential landscape component to maintain the function of natural systems (see Chapter 5 for analysis of this planning tool). Corridors provide landscape connectivity for wildlife movement and provide stepping-stones to keep species matrices intact and functioning (Noss, 1983; Van Lier and Cook, 1994). They allow seasonal movement for feeding, population dispersal and support of metapopulations; they also prevent inbreeding and help maintain genetic variability within a population. River systems are examples of natural corridors, which maintain aquatic conditions such as water temperature and oxygen content.

One of the major problems facing natural resource and land use planners is that urban development blocks or disrupts natural corridors. Fragmenting corridors can: 1) reduce area of habitat available to species; 2) increase likelihood of population extinction by limiting immigration; and 3) exacerbate genetic problems resulting from inbreeding. For these reasons, high quality environmental plans seek to protect and/or create corridors to facilitate connectivity.

Third, corridors interconnect to form *networks* that enclose other landscape elements. Ecological networks create connectivity and enable the functioning of landscapes by allowing flow of species and energy. In this way networks provide an ecological structure or foundation for multi-species matrices (Dramstead et al., 1996). Van Langevelde (1994) suggests that habitat networks may be essential to the survival of populations of native species. They provide opportunities for efficient migratory routes and alter the flow of nutrients, water and energy across landscapes. Protection of this "ecological infrastructure," the constellation of landscape elements that is functional for the dispersal of a species in a landscape, is therefore critical to protecting ecosystems over the long term (Van Lier and Cook, 1994).

Fourth, protecting the "landscape mosaic" through connectivity is considered important to protecting overall biodiversity because it safeguards *metapopulations*. This concept, stemming from island biogeography theory, is the building block for species diversity. A metapopulation is commonly defined as any set of spatially defined local populations, which are demographically affected by the spatial arrangement of habitat patches and the resistance of the non-habitat of the landscape matrix (Van Langevelde, 1994). The survival of metapopulations is based on the size of the habitat matrix, its isolation or degree of connection to other habitats, and the level of human disturbance to the habitat. Removal of a habitat related to a metapopulation (for example through commercial development) reduces its size and increases the probability of local within habitat extinctions. Figure 6.1 illustrates the spatial structure of a metapopulation of deer. Deer habitats are connected across space through movement and dispersal. Local extinctions in one habitat reduces the overall size of the metapopulation, provides less habitat options for deer, and may reduce the survival probability of the total population.

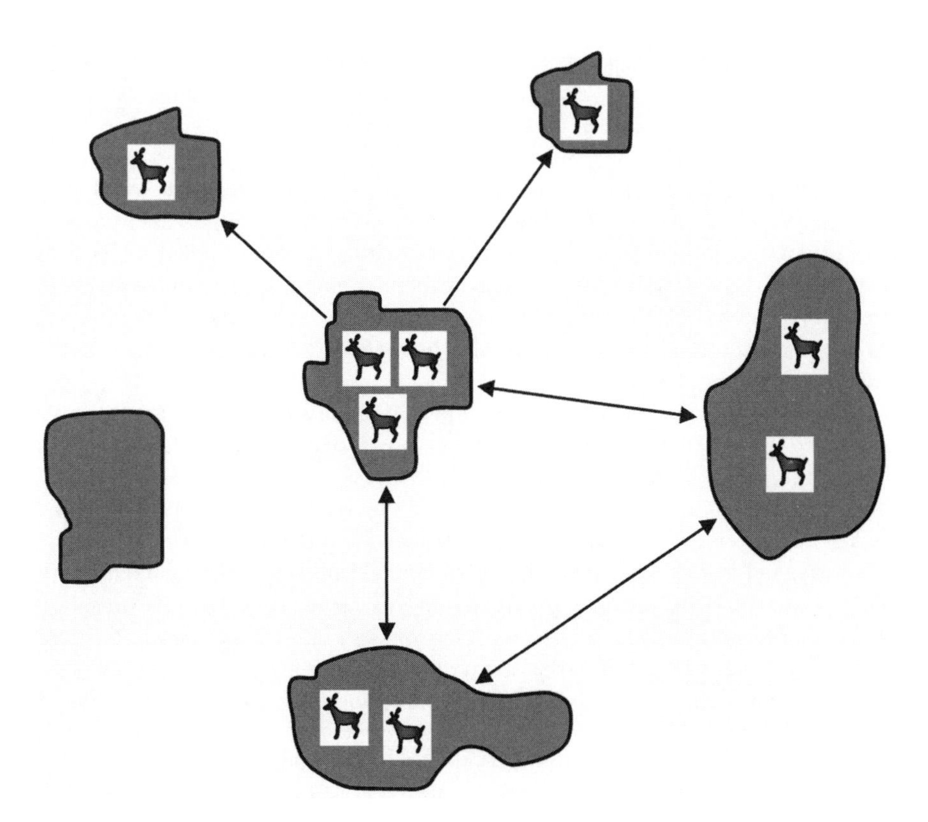

Figure 6.1 Spatial structure of metapopulations
Source: Adapted from Peck, 1998.

One recent and important application of ecological network analysis theory is the establishment of a system of protected areas. A network design can effectively mimic spatial linkages or migratory patterns over a broad region and protect species in critical life-history phases (Ray, 1996). While protected area network theory initially came from the terrestrial sciences with the key conceptual works of Harris (1984), Noss and Cooperrider (1994), and Jongman (2004), the establishment of multiple protected areas is particularly relevant in the coastal and marine environment, where habitats are more connected than terrestrial environments due to the constant movement of water. A coherent network or system of protected areas could provide a more effective means of protecting transboundary resources by better addressing issues contributing to the decline of water-influenced terrestrial and marine ecosystems that span large areas (Shafer, 1995; Murry et al., 1999).

By accounting for habitat linkages, biogeographic processes, and the constant flow of resources through the system, a network approach can more effectively achieve commonly held conservation and management goals. The basic elements of reserve design include core areas to protect critical habitats, corridors to link core areas, and buffer zones adjacent to the cores or corridors for additional protection. Establishing networks of protected areas provides one of the strongest approaches to protecting ecosystem integrity and can be applied at a variety of spatial scales. While the practice is slow to gain acceptance by environmental planners, if ecosystems are to be managed over the long term, this concept and its applications must become a central goal of local natural resource planning programs in the US.

One method used for developing overall frameworks for biodiversity conservation at large spatial scales by applying the theory of a network of protected areas is called Gap Analysis (Scott et al., 1993). This method, based on the interpretation of remote sensing data and the use of Geographic Information Systems (GIS) to identify areas of biodiversity, provides the strongest methodology for applying ecosystem management concepts to actual ecosystem problems. Its replication across the US has generated much of the data needed by planners to protect and manage the integrity of ecological systems.

Gap Analysis relies on GIS to identify gaps in biodiversity protection that may be filled by the establishment of new preserves or changes in land use practices (Davis et al., 1990). Gap Analysis uses the distribution of actual vegetation types and vertebrate (as well as butterfly species) as indicators or surrogates for biodiversity. Digital map overlays in a GIS are used to identify individual species, species-rich areas, and vegetation types that are underrepresented in existing biodiversity management areas (Edwards et al., 1993). In this sense, Gap Analysis organizes existing survey information to identify areas of high biodiversity before they are degraded. It functions as a first-pass approach for organizing biological information and is not a substitute for a detailed biological inventory. Depending on the nature of the issue, the database can be used to springboard into other, more detailed studies and is meant to be employed as a proactive rather than reactionary management tool (Scott et al., 1991). Because of their usefulness in identifying areas of high biodiversity across landscapes, Gap Analysis data are used to determine levels of biodiversity in the study area. Florida completed a state-wide Gap Analysis in 1994.

Table 6.1 Landscape ecology principles and planning tools

Landscape ecology principles	Landscape planning design principles	Landscape planning tools
Connectivity/corridors	Maintain ample interior habitat (large patches over small)	Establishment of protected areas/reserves
Multiple spatial/temporal scales	Reduce edge habitat	Fencing controls
Disturbances	Circular reserves over elongated	Restrict removal of native vegetation
Habitat patches	Establish corridors between habitat patches	Remove exotic species and prevent their introduction
Habitat edges	One large patch is better than several smaller ones (in some cases)	Buffer habitat patches and sensitive areas from high impact uses
Constant change/dynamics	Create separate patches close together	Density restrictions or land use restrictions adjacent to or within patches
Landscape matrix/structure	A triangular design of separate patches is better than a linear design	Establish conservation overlay zones
Networks	Maintain viable native populations	Use of transfer development rights away from intact patches
Metapopulations	Reduce physical barriers	Density bonuses to direct development away from critical patches
Dispersal	Establish habitat patches as stepping stones to maintain ecological infrastructure (reduce isolation of habitat)	Controls on construction activities (e.g. dredging, excavating, roadways, vegetation removal)
Genetic drift	Stepping stones should have alternative routes or loops for dispersal	Use of land acquisition programs (e.g. conservation easements, fee simple purchase, etc.)
Species/community interaction	Create complex/diverse/ curvilinear edges (rather than smooth or abrupt)	Controls on vehicular access to sensitive habitats
Movement of energy, species, etc.	Create convoluted over round patches	Capital improvements programming to direct growth away from intact patches

Table 6.1 continued

Landscape ecology principles	Landscape planning design principles	Landscape planning tools
Ecosystem concepts	Elongated patches should be perpendicular to dispersing individuals	Clustering development to maintain large or connected patches
	Corridors should be wide and diverse in composition	Monitoring of landscape mosaic and human impacts
	Use river corridors for dispersion	Preferential tax treatments for developing away from native patches
	Establish windbreaks as natural barriers	Restoration of previously impacts habitat patches
	Reduce fragmentation of previously connected patches	Use of GIS analysis to identify and reduce fragmentation of native patches
	Create convergence points, adjacency, and interspersion of habitats	Education programs
	Concentrate high impact uses	Targeted growth areas
	Prevent the spread of exotic species	Phasing of development
	Integration of land uses over segregation	
	Protect indicator species	
	Increase the overall number of habitat patches	

In summary, the concepts underlying landscape ecology and conservation biology provide a scientifically defensible framework for managing ecosystems over the long term. They can help decision makers understand the meaning of biodiversity and the necessity to focus management efforts on ecological rather than political or administrative boundaries. Furthermore, the literature on this topic provides clear goals for protecting ecosystems, such as maintaining connections among wildlife habitats or maintaining large intact patches of native species by preventing fragmentation of those patches by development.

The literature, however, does not effectively connect ecological principles with the practice of land use planning. Foreman and others go as far as generating "design principles" based on ecological concepts, but fall short when it comes to converting these principles into planning tools. Table 6.1 provides a comprehensive list of landscape ecology and design principles along- side landscape planning tools and strategies. Linking land use planning tools to ecological concepts helps form a more complete understanding of how a plan can manage ecological systems. Since a major goal of a plan focused on the management of ecological systems or the conservation of critical natural resources is to protect biodiversity and its landscape components, specific policies will depend on the type and condition of existing habitats. For example, a jurisdiction with intense urban development and fragmented critical habitats will develop a plan very different from a jurisdiction with a more complete existing landscape mosaic. Existing conditions related to the health of metapopulations, the degree of connectivity across the landscape, and the maintenance of keystone species will all influence the focus and content of a local plan.

The Effects of Biodiversity and Human Disturbance on Plan Quality

Given the importance of biodiversity in supporting viable ecosystems and the increasing emphasis on protecting biodiversity in environmental plans, it seems logical that planners and stakeholders involved in drafting plans will be influenced by the amount of biodiversity contained within their community. As proactive policy statements, the environmental elements of comprehensive plans identify existing critical natural resources, recognize their value, and seek to protect these resources for future generations. Thus, as a major factor influencing conservation and management efforts, it is postulated that the amount of biodiversity in a jurisdiction will have a positive impact on the quality of management plans and strategies (Noss and Scott, 1997; Peck, 1998).

Higher levels of biodiversity may increase local ecosystem plan quality because there will be a greater perceived need to protect valuable natural resources before they are irreversibly damaged. Since the purpose of comprehensive plans is to act as long-range policy instruments, conservation elements should take a precautionary stance when it comes to sustainable resource management. Jurisdictions with high biodiversity should be interested in safeguarding critical ecological components with directed goals and policies for future generations (Kirklin, 1995).

However, with reduced levels of biodiversity, planners and planning participants may feel an urgency to protect natural resources, which will in turn increase ecosystem plan quality. Levels of biodiversity then, are intricately connected to levels of disturbance within a landscape. Since ecosystem management efforts are often reactions to some level of environmental crises (e.g. loss of seagrass in the Chesapeake Bay, water quality declines in the Everglades, loss of the Spotted Owl in the Northwest, etc.), human threats to biodiversity or disturbance to habitat may also positively impact plan quality (Wondolleck and Yaffee, 2000). Human disturbance to habitat occurs in many forms, but is mostly driven by increased impervious surfaces associated with urban development, loss of native vegetation from forestry and agriculture, the introduction of exotic or invasive species into a native ecosystem, and water pollution caused by urban run-off. Under this notion, the higher the perceived (or actual) degree of threat, the stronger the expected level of plan quality.

Reactionary approaches to environmental planning are not entirely new phenomena. Over 20 years ago, Burby and French (1981) discovered a similar policy response they termed a "land use management paradox." In their study, communities tended to enact strong hazard management programs only after the damage to or development of the flood zone had taken place. Hazard mitigation strategies were installed as reactionary strategies rather than proactive measures to avert loss of critical natural resources and, in this case, human life. The paradox emerges because communities protected their flood plains once development had already taken place, causing these policies to be far less useful in accomplishing planning goals. Although this study used different variables, measurements, and analyses, the same type of paradox applies to the amount of biodiversity or critical habitat within a jurisdiction and corresponding efforts at ecosystem planning. In these cases, communities may implement goals, policies, and strategies to protect ecosystem integrity only when there is little left to protect. Rapid human growth and development resulting in disturbance under this hypothesis will drive ecosystem plan quality.

These instances have become known as "train wrecks" throughout the environmental policy community (Haeubner, 1998). "Train wrecks" occur when there are clashes between urban development and biodiversity, which spur major environmental initiatives such as the protection of the spotted owl in the Northwest or the attempted restoration of the Everglades in south Florida. While these "wrecks" could have been avoided with sound planning, they were seen as necessary to bring about environmental efforts in the first place.

Ruth (1990) captures this environmental planning problem in her description of two philosophies or approaches of natural resource managers: 1) damage control, and 2) anticipation/prevention. Damage control-driven planning and management reacts to negative criticism and clearly demonstrated problems. Ruth (1990) terms this outdated approach a dinosaur because it reacts to problems rather than anticipating and preventing them. In contrast, management propelled by anticipation/prevention proactively resolves environmental conflicts before they become intractable.

Including human disturbance in a conceptual model is not enough to isolate the effect of disturbance in relation to other environmental factors on local ecosystem plan quality. As discussed above, a conceptual model must consider that disturbance and biodiversity are intricately linked concepts and measures. Increasing levels of disturbance will invariably result in decreasing levels of biodiversity. Although human disturbance on natural ecosystems may alone stimulate the adoption of higher scoring plans, if that disturbance is also associated with the loss of high biodiversity, the motivation to enact environmental plans may be even greater. A perceived environmental problem or threat, such as habitat loss most often initiates the adoption of environmental plans (Lein, 2003). Increasing attention to and awareness of the problem can help open a "policy window" of opportunity to generate plans to mitigate continued decline of ecosystem components (Kingdon, 1984; Haeubner, 1998). For this reason, the impact of biodiversity on plan quality may be dependent on the level of disturbance. Disturbed-biodiversity may have the largest impact on ecosystem plan quality and therefore must be included in a model explaining ecosystem plan quality as the interaction between biodiversity and disturbance.

Parallel research in environmental and natural hazards mitigation further illustrates the relationship between the perception of threat and policy change. Focusing events help generate public interest in a particular problem and trigger the policy making process (Birkland, 1997). Increased attention based on the perceived seriousness of the problem is thus an essential precondition for action (Turner, 1986; Lindell and Perry, 1999). For example, Lindell and Prater (2000) found that the level of personal intrusiveness of a seismic event (based on the frequency a respondent thought and talked about an earthquake) is a significant predictor of seismic hazard adjustment. They observed that when the perception of threat is heightened, it is more likely to be addressed by taking action.

Chapter 7

The Role of Biodiversity in Local Plan Making

This chapter presents the results of two empirical studies focusing on the degree to which biodiversity influences local jurisdictions in Florida to adopt high quality environmental plans. The first case measures the amount of biodiversity within local jurisdictions and models its impact on the quality of comprehensive plans associated with ecosystem management. The second case uses GIS to map and measure the policies of neighboring jurisdictions in the southern part of Florida and better understand how biodiversity and human disturbance of critical natural resources affects the quality of these policies.

CASE 1: Examining the Effects of Biodiversity on the Ability of Plans to Manage Ecological Systems

This case study is the first of several to identify factors influencing the quality of local plans. Specifically, it seeks to form a better understanding of how local jurisdictions respond to declining levels of critical natural resources by explaining how the quality of these plans is influenced by the amount of biodiversity and the degree of threat placed on the existing natural resource base within local jurisdictions. By investigating the effects of biodiversity and human disturbance of this biodiversity on our measure of ecosystem plan quality, this case tests and confirms the land use management paradox presented in the previous chapter where communities adopt environmental plan components only after much of the critical natural resources they intend to protect are lost to human development.

In addition to variables for biodiversity and disturbance, a series of contextual control variables were included in a statistical model to further identify the importance of environmental variables to plan quality. Population (Berke et al., 1998), wealth (Berke et al., 1996), planning capacity (Burby and May, 1998), and agency commitment (Berke et al., 1996) have all been shown to have positive effects on various measures of plan quality. Jurisdictions with larger populations usually have more complex environmental problems that result in a need for strong planning. Wealthier populations usually have more financial resources to devote to planning staffs and plan development. The higher the planning agency capacity for a given jurisdiction, the more technical expertise and personnel devoted to producing the plan. Finally, agency commitment to critical habitat protection should positively influence plan quality by emphasizing the importance of habitat protection and devoting time during the planning process to discuss pertinent environmental issues.

The Sample

The same random sample of 30 local jurisdictions generated for Case 1 in Chapter 5 was replicated for this study to isolate the influence of biodiversity on ecosystem plan quality. In addition, the same five component plan coding protocol and measurement procedure was used to derive an overall score for ecosystem plan quality based on a scale of 0-50 (for more detail see Case 1, Chapter 5).

Measuring Biodiversity and Disturbance

Satellite images of land cover generated by the Florida Fish and Wildlife Conservation Commission (FFWCC) were used to predict species overlap and identify "hot spots" of biodiversity. Areas of biodiversity based on the overlap of 44 focal species were selected for final analysis, since they consider the broadest biological factors over both public and private lands. Focal species serve as umbrella or indicator species of overall biodiversity in Florida (Cox et al., 1994). Each pixel in the raster-based data layer was assigned a value on a scale of 1-3 depending on the number of species overlap. The amount of biodiversity was measured by calculating the area of all values (1-3) and dividing that value by the total acreage of a jurisdiction so that the variable could be interpreted on a scale of 0-1.

The amount of disturbance was calculated in a similar manner based on the same land cover image developed by the FFWCC. Areas interpreted as disturbed land cover (grassland and agriculture, shrub and brush, barren and urban, and exotic species) were summed in a rasterized coverage and then divided by the area of a local jurisdiction creating a disturbance variable on a scale of 0-1. Disturbed-biodiversity was measured as simply the interaction of biodiversity and disturbance.[1]

Does Biodiversity Matter?

Statistically testing the propositions set forth in Chapter 6 reveals a finding which for some may be intuitive and for others quite surprising.[2] On average, the proportion of area with high biodiversity within a jurisdiction has no significant statistical bearing on plan quality (in fact the coefficient is negative). However, the area of biodiversity that is associated with disturbance generates markedly higher quality plans (Table 7.1).

---

1 The means of biodiversity and disturbance were subtracted before the interaction was performed. This commonly performed statistical procedure reduces the threat of multicollinearity in the model (Akin et al., 1991).

2 Analyses of the data were based on Ordinary Least Squares (OLS) regression. Several statistical tests for reliability were conducted to ensure the OLS estimators were Best Linear Unbiased Estimates (BLUE). Tests for model specification, multicollinearity, and heteroskedasticity revealed no violation of regression assumptions. In addition, a series of diagnostics was performed to test for influential data points or outliers in the data set. Given the small sample size, influential data points may have a significant impact on the interpretation of ecosystem plan quality. Various types of plots, as well as robust regression, uncovered no influential data points affecting the results.

Disturbance by itself is also a significant factor (where $p<.05$) in raising the quality of plans in the sample. These results support the hypothesis that increasing levels of disturbance or threats to biodiversity will result in higher quality local comprehensive plans. In other words, an increased proportion of human disturbance, such as pavement, agricultural practices, and the presence of invasive species within a jurisdiction, is a major environmental factor driving ecosystem plan quality as measured in this book. Only when biodiversity or critical habitat is under threat from anthropogenic stresses (e.g. urban development) does it appear to have a significant positive impact on plan quality.

Table 7.1 The impact of environmental variables on plan quality[a]

Variable	Coefficient	Standardized coefficient	Standard error	T-value	P-value
Area of jurisdiction with biodiversity	4.74	.077	11.33	.419	.68
Area of jurisdiction with disturbance	13.05	.386	4.66	2.801	.013
Disturbed-biodiversity	139.95	.469	47.60	2.94	.010
Population[b]	4.79	.382	1.77	2.70	.013
Wealth[c]	10.26	.207	4.92	2.088	.049
Capacity[d]	.0071	.003	.266	0.027	.979
Commitment[e]	2.09	.166	1.431	1.460	.164
Constant	9.24		3.02	3.063	.005
N:	30				
F-Ratio (7,22):	17.03				
Significance:	.0000				
Adjusted R-squared:	.795				

Source: Adapted from Brody, 2003: 827.

Note: [a] plan quality is the total plan coding score divided by the total possible score and multiplied by 10 to create a scale from 0-50; [b] population is the natural log of US Census population estimates for 1997; [c] wealth is the natural log of US Census estimates of median home value; [d] capacity is the number of planners involved in developing the plan; [e] commitment is the degree of effort spent on the issue by the local government combined with the degree to which the government emphasized the issue during the planning process.

This case study suggests that planners and planning participants developing comprehensive plans are reacting to the degradation of critical natural resources and are driven by the incidence of environmental "train wrecks" to generate high quality ecosystem-based plans. On the other hand, with high levels of undisturbed biodiversity, there seems to be less of a perceived need to protect critical natural resources within the context of comprehensive planning. Without the warning signals of habitat fragmentation and loss of keystone species, planners seem to lack motivation to initiate early protection measures.

Take, for example, Pinellas County. With only 280 square-miles, Pinellas is the second smallest county in the Florida. Its small land area and comparatively large population make it the most densely populated county in the state with 3,228 persons per square-mile. As a result, less than 10 percent of the County is considered vacant and available for urban development. Rapid growth and development from the 1950s through the 1980s led to a reactionary interest in environmental planning (Brody, 2001). At present, Pinellas is approximately 92 percent urbanized, and as the County approaches a completely built-out stage, its government and community are focused on protecting remaining pockets of open space and wildlife habitat. As a consequence, the Pinellas County 1998 comprehensive plan is extremely strong in terms of protecting the integrity and function of ecological systems both within and adjacent to its borders. In fact, its total plan quality score is the highest in the study sample (see Case 1, Chapter 5). The commitment to the protection of biodiversity and ecosystem management has emerged in the Pinellas County plan after most of the urban and suburban development had already taken place. While strong goals and policies are set in place, there is relatively little remaining to protect and management in the way of critical natural resources.

Even after adding contextual controls, local jurisdictions associated with anthropogenic disturbance of biodiversity remain the most powerful predictors of local ecosystem plan quality. While disturbance-related variables remain statistically significant, there is a noticeable increase in the p-values compared to the initial analysis of environmental variables. This decrease in significance may be associated with the inclusion of population in the model, which has a significantly positive impact on ecosystem plan quality. Population can often be associated with increased urban development and decline of critical habitats or overall biodiversity. Growth pressures are associated with higher levels of disturbance to habitat, resulting in a greater perceived need to protect remaining areas of biodiversity. The addition of population thus causes some redundancies in measurement (as evidenced by a high zero-order correlation between population and human disturbance) that may account for the decrease in significance of some environmental variables.

Wealth, as measured by the medium home value within a jurisdiction, is also a significant factor in explaining ecosystem plan quality. Jurisdictions with wealthier populations usually have more financial resources to devote to planning staffs and plan development, leading to the adoption of higher quality plans. Furthermore, residents with high incomes are also often more educated and have more time and interest in participating in the planning process, particularly when it comes to environmental issues. These two factors may explain the significant positive effect of wealth on ecosystem plan quality.

Perhaps the most salient result is the significance of the interaction of biodiversity and human disturbance where disturbance to biodiversity drives ecosystem plan quality significantly higher. This interaction was investigated in more detail by observing the impact of disturbance on ecosystem plan quality when biodiversity was set at different levels. Significance levels for disturbance were calculated for plan quality when biodiversity was set at its minimum, mean, and maximum

(Table 7.2). In terms of significance levels, disturbance has the greatest effect on the dependent variable when biodiversity is at its extremes. Human disturbance may be most noticeable to planners and planning participants when the amount of biodiversity is either very low or very high. Even more insightful, however, is the dramatic increase in the coefficient of disturbance as levels of biodiversity increase. When biodiversity is at its maximum value, the effect of disturbance on ecosystem plan quality is extremely strong. This finding further supports the proposition that the combination of high biodiversity and disturbance is the most powerful predictor of ecosystem plan quality.

Table 7.2 Interaction of biodiversity and disturbance

Biodiversity level	Disturbance level	
	Coefficient[a]	P-value[b]
Minimum	9.63	.030
Mean	17.44	.10
Maximum	46.64	.038

Source: Adapted from Brody, 2003: 829.

Note: [a] the covariance between the parameter estimates of disturbance and plan quality when biodiversity is set at a specific value; [b] level of significance of disturbance on plan quality when biodiversity is set at a specific value.

Can Planning Be Proactive?

The most significant finding of the above analysis shows that the degree of disturbance or threat to biodiversity is the strongest predictor of ecosystem plan quality. While this finding may be obvious to some, it is not good news for the environmental planning profession. Even though comprehensive planning is intended to be a proactive policy-making process where communities lay out their long-term vision of the future, the quality of the plans increases only after there is a clear and present loss of biodiversity. Some degree of adverse impact to critical natural resources can be productive in manifesting an environmental problem, thereby generating interest in ecological management and producing high quality plans. However, this study confirms the "land use management paradox" by finding that planners and planning participants are reacting to the loss of biodiversity at the point where there is little left to protect. The threshold for planning response in Florida appears to be so high that the integration of ecosystem management abilities at the local level is essentially counter-productive (Brody, 2003). A "damage-control" approach to natural resource management must rely on restoration activities. This approach to environmental planning is costly, inefficient, and in many instances practically infeasible.

Because local jurisdictions can greatly impact ecological systems and their components through land use decisions, increasing the ability of land use plans to manage entire natural systems rather than a fragment is critical to attaining state and federal environmental goals. The central issue for local ecosystem planning thus is determining how to motivate communities to protect critical ecosystem components *before* they are lost to human growth and development. Motivating action involves increasing the sensitivity of the planning response threshold so that those involved in drafting a plan are stimulated to protect ecosystem components early in the process of natural resource decline. The general conclusion that replacing areas of biodiversity with pavement is the most effective way to produce high quality plans is not a productive alternative for local planners.

While further study is needed to understand how to lower the environmental planning response threshold, there are several recommendations that may help communities in Florida and across the US incorporate ecosystem considerations into plans and planning processes before substantial degradation of biodiversity takes place. These recommendations are described in more detail in Chapter 12 of this book.

First, monitoring activities can be an essential proactive planning lever for ecosystem management. Updated information on the conditions of critical habitats and ecological processes can help planners and local officials adopt policies that buffer potential catastrophic decline. A second proactive planning practice involves the use of Geographic Information Systems (GIS). GIS techniques help visualize anticipated impacts from growth and development decisions. Since spatial data is becoming ever more abundant and these software programs are becoming more affordable and easier to use, the use of GIS is a viable option even for small, rural communities without vast resources.

A third potential proactive planning lever is the use of incentive-based policies and programs. The plans examined in this case study generally do not emphasize incentive-based tools or policies. Instead, jurisdictions concentrate primarily on a narrow set of regulatory actions, such as land use restrictions or conservation zoning. However, the use of incentive-based policies, such as density bonuses, transfer of development rights, and preferential tax treatments (see planning protocol in Chapter 4) can effectively achieve the goals of ecosystem management at the local level by encouraging rather than forcing property owners to protect critical habitats and areas of high biodiversity. Finally, environmental education programs are one of the most powerful ways to foster proactive ecosystem management practices. Local outreach, such as workshops, printed and electronic information, and community presentations can build public awareness on the importance of protecting the value of critical natural resources and maintaining ecological integrity. Only half of the sample analyzed in this case included public environmental education programs in its set of policies, indicating that the link between planning and education is being underemphasized in Florida.

CASE 2: Mapping the Collective Capabilities of Local Jurisdictions to Plan for Ecological Systems

While a large amount of research suggests collaborative ecosystem management is a desirable approach to protect the integrity of critical natural resources in the US, comparatively little work has been done to show how local jurisdictions are playing a role in the management of large-scale natural systems. This case study builds on Case 2 in Chapter 5 by evaluating the collective potential of local jurisdictions to manage transboundary ecological systems (as defined by watershed units). Specifically, it uses GIS to map, measure, and analyse the existing mosaic of management based on policies in local comprehensive plans across 22 adjacent watersheds in southern Florida. In addition to describing the spatial pattern of watershed plan scores based on local level plans and policies, we seek to explain the major factors contributing to the strength or weakness of local jurisdictions to manage transboundary ecological systems.

The Sample

Watersheds have been identified as an ideal planning unit for ecosystem managers when considering the protection of ecological processes and critical natural habitats (Williams et al., 1997). We selected twenty-two adjacent watersheds for analysis in the southern portion of Florida defined by the United States Geological Service's (USGS) fourth order Hydrological Unit Code (HUC). In areas south of Lake Okeechobee, we took direction from Florida's Department of Environmental Protection (DEP) which redefined watershed boundaries due to human alteration and fragmentation of traditional water flows. The sample of watersheds stretches from the west coast near Tampa Bay to the heavily developed southeast coast of the State, representing a wide variety of biophysical regions and institutional/political settings (see Figure 7.1).

Local jurisdictions were then selected containing land area within one of the twenty-two watersheds. Thirty adjacent counties intersecting the watershed boundaries, plus the fifteen largest cities in land area were selected for analysis (Figure 7.2). Since our goal is to achieve the greatest level of spatial coverage, cities were selected based on area rather than by population size. Watersheds in our sample contain an average of 5.13 jurisdictions. As was done in Case 2 in Chapter 5, the most recent comprehensive plans for these counties and cities were evaluated against the ecosystem plan quality protocol (presented in Chapter 4) containing indicators to determine their collective ability to manage watersheds or, more generally, ecological systems.

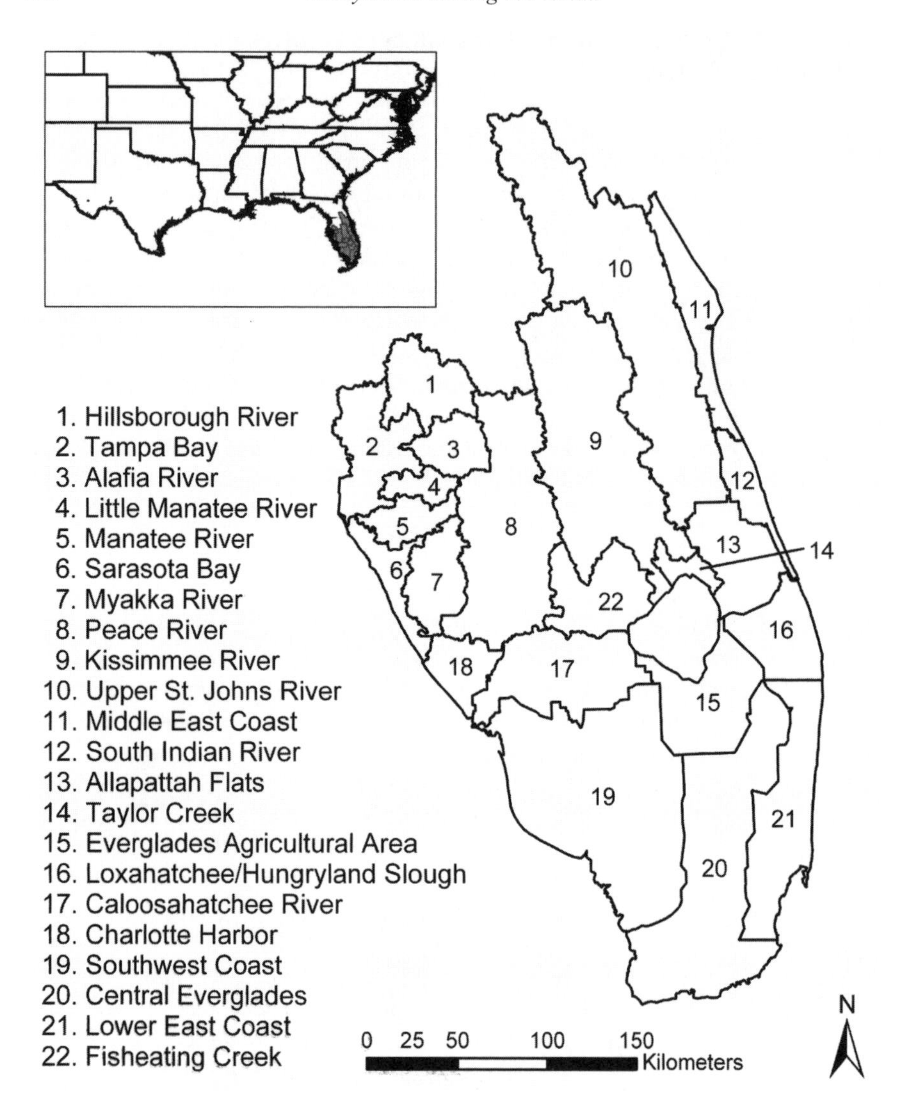

Figure 7.1 Selected watersheds
Source: Brody et al., 2004: 40.

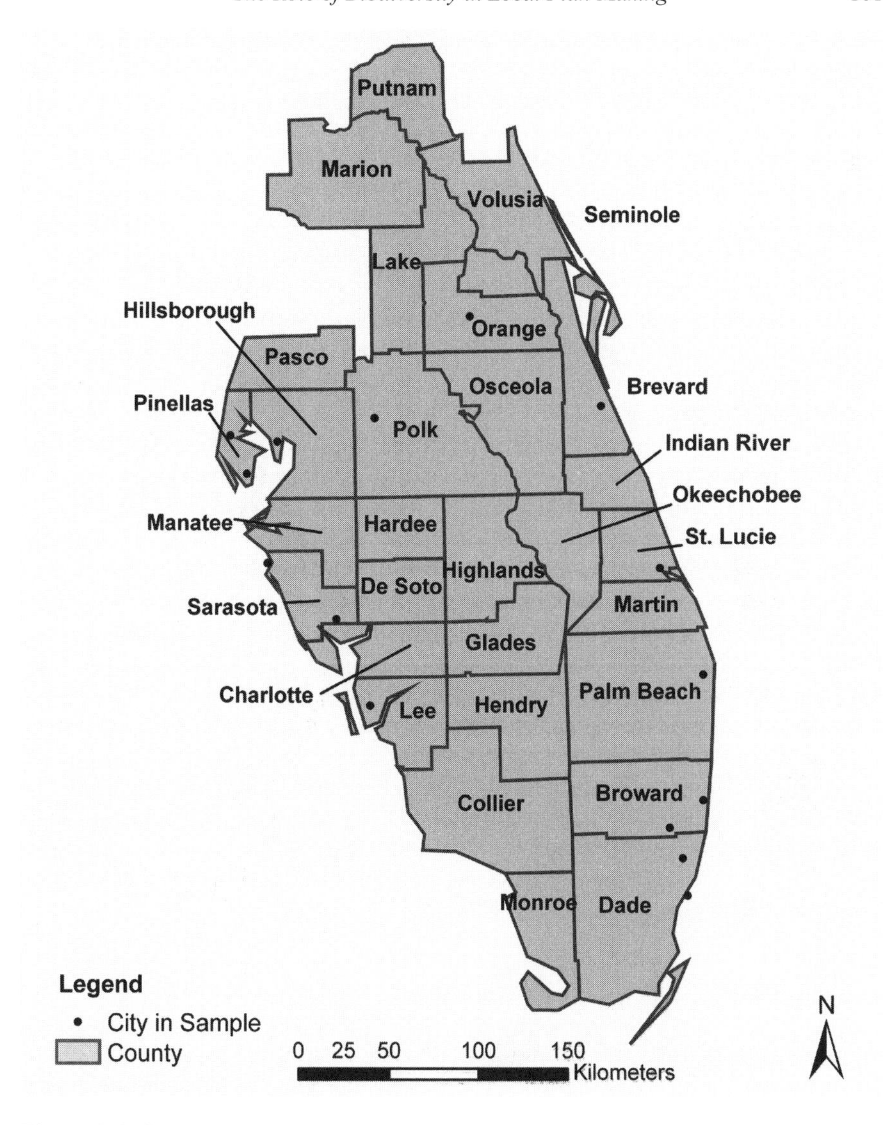

Figure 7.2 Selected local jurisdictions
Source: Brody et al., 2004: 41.

Measuring Collective Planning Capabilities

Ecosystem plan quality was measured by evaluating the most recent comprehensive plan (as of 2004) for each jurisdiction located within a selected watershed against the 123 ecosystem management indicators (see Table 4.1 for a complete listing of indicators comprising the coding protocol). The same scoring procedure used for Case 1 of Chapter 5 (refer to this Chapter for further details) was replicated here so that each indicator was measured on a 0-2 ordinal scale. Measures of overall ecosystem plan quality were also calculated using the same procedure in Case 1 of Chapter 5, creating indices for each plan component (on a scale of 0-10) and overall plan quality (on a scale of 0-50). We followed the procedure used in Case 2 of Chapter 5 to measure and map plan quality scores based on a jurisdiction's area within each watershed (as opposed to Ecosystem Management Areas which served as the unit of analysis in Case 2, Chapter 5). Briefly, we used GIS to measure the percentage of spatial coverage for each plan quality component within each watershed. We then used this proportion to weight that jurisdiction's contribution to the watershed score on that plan component (PC_j) and the total plan quality score (TPQ_i).

We measured ecological and human disturbance variables using the same method as those used in Case 1 above. We calculated the amount of biodiversity and disturbance by calculating the area of all values associated with the satellite images and dividing that value by the total area of a watershed (see Case 1 above for more details). Similar to the previous case in this chapter, contextual variables are also important predictors of ecosystem plan quality. We measured socioeconomic and demographic independent variables with data obtained from the 2000 US Census. Population, population growth between 1990 and 2000, wealth (median home value), education (percentage of the population with a high school degree), and land use were recorded for each jurisdiction occupying a selected watershed. Land use was measured based on five different types of uses which include commercial, industrial, agricultural, multi-family residential, and single-family residential. We then weighted the values of each variable by the proportion (P_{ij}) of area in a watershed that was occupied by that jurisdiction. Finally, we calculated ecosystem level variables by summing all weighted values within each watershed. Additional influencing factors include the number of jurisdictions within each watershed, the area of each watershed (as calculated by the GIS program), and planning agency capacity. Information on planning capacity was obtained by contacting each planning department in the sample and measured based on the number of staff devoted to writing the comprehensive plan.

Assessing Plan Quality Scores by Watersheds

Our first step in this analysis is to provide a statistical and graphic assessment of the degree to which local jurisdictions are collectively managing watersheds in southern Florida based on an assessment of their local comprehensive plans. As reported in Table 7.3, the mean score for total ecosystem plan quality is 18.43, which on a scale of 0-50, indicates a relatively weak potential effort to manage ecological systems at the local level.[3] Mean scores for all plan components (scale of 0-10) also register fairly low despite a federal initiative to restore and manage the Everglades system, a strong state program on ecosystem and regional watershed management, and a prescriptive local comprehensive planning mandate, which entails protecting critical habitats and ecological functions.

3 Note this score is consistent with the random sample of plans evaluated against the same plan coding protocol evaluated in Case 1, Chapter 5.

Table 7.3 Plan quality scores by watershed

Watershed	**Factual Base**	**Goals and Objectives**	**Coordination Capabilities**	**Policy**	**Implementation**	**Total Plan Quality**
Alafia River	3.02	3.66	3.11	2.93	1.18	15.11
Allahpattah Flats	4.01	5.93	4.95	5.12	3.41	24.31
Caloosahatchee River	2.77	2.39	3.95	4.23	2.21	17.10
Central Everglades	1.46	4.08	4.94	4.17	4.30	20.04
Charlotte Harbor	3.15	2.70	3.90	3.87	1.56	16.37
Middle East Coast	0.80	2.58	3.66	3.42	3.40	14.02
Everglades Agricultural Area	1.91	5.56	7.08	6.97	6.57	29.30
Fisheating Creek	3.42	2.60	3.17	4.13	2.05	15.93
Hillsborough River	3.11	3.85	3.67	4.51	0.80	17.00
South Indian River	3.79	2.93	4.00	4.02	3.86	18.80
Kissimmee River	2.78	2.43	4.19	3.27	2.29	15.48
Little Manatee River	3.43	4.85	3.74	4.49	1.96	20.12
Lower East Coast	1.10	4.68	5.70	4.75	4.55	21.81
Loxahatchee/ Hungryland Slough	2.72	6.96	7.45	7.42	8.02	33.55
Manatee River	0.57	5.03	6.26	5.65	4.70	23.17
Myakka River	3.16	3.77	4.49	5.05	2.03	19.50
Peace River	1.08	2.69	3.41	2.83	1.91	12.46
Sarasota Bay	4.03	2.62	3.43	3.94	0.83	15.94
Southwest Coast	1.58	2.58	4.37	4.15	2.44	16.57
Upper St Johns River	2.36	3.97	5.02	4.76	4.17	20.87
Tampa Bay	3.83	4.20	3.88	4.16	3.16	21.00
Taylor Creek	1.17	2.64	4.08	2.27	3.24	13.60
Average Score	**2.41**	**3.61**	**4.29**	**4.19**	**3.00**	**18.43**

Source: Adapted from Brody et al., 2004: 43.

The *Factual Basis* is the lowest scoring plan component, demonstrating a general lack of local knowledge regarding existing natural resources, human impacts to these resources, and their management status within a given watershed. This finding is consequential since goals, objectives, and policies rely on a thorough understanding and inventory of the natural resource to be managed by the plan. The highest scoring watersheds are for the most part associated with high profile bays located in the western portion of the state (see Figure 7.3). Allahpattah Flats is a notable exception where the jurisdictions within this watershed seem to collectively make the effort to catalogue and analyze their natural resource base. Various ecological surveys and studies of human impacts to water quality have been conducted in Tampa and Sarasota Bay partly because of their designation as estuaries of national importance under EPA's National Estuary Program (NEP). Increasing interest in understanding these natural systems and the availability of resources to conduct studies most likely contributes to the high scoring factual basis of the local plans associated with these two watersheds. In contrast, watersheds with the lowest scoring factual basis tend to be inland (Little Manatee River) or coastal (East Coast Middle) areas that receive less attention for their ecological importance and are not considered high priorities for ecological study.

Scores for the *Goals and Objectives* and *Policies, Tools and Strategies* plan components follow a similar trend. Both plan components score highest for watersheds to the west and directly south of Lake Okeechobee. Allahpattah Flats, Loxahatchee/Hungryland Slough, and the Everglades Agricultural Area are among the highest scoring watersheds. These ecological units, surrounding the Everglades region, contain some of the fastest growing and most planning-oriented local jurisdictions in Florida. The plans of communities located within high scoring watersheds include both broad goals and specific measurable objectives for managing watershed systems. Associated policies tend to be mandatory and include not only traditional regulatory measures, but also incentive-based and other non-regulatory tools. Surprisingly, Sarasota Bay, which is among the top scoring watersheds for its *Factual Basis*, is not as strong when it comes to goals, objectives, and policies associated with ecosystem approaches to management. Plans with strong factual basis often build upon this foundation with well-crafted environmental strategies.

The *Inter-jurisdictional Coordination and Capabilities* plan component is, overall, the highest scoring of the five plan components (4.3 on a scale of 0-10). This result suggests that jurisdictions recognize the transboundary nature of ecosystems and are committed to collaborating with other jurisdictions to manage these natural resources over the long term. Because this study evaluates watersheds crossing multiple jurisdictions, collaboration is an essential component for effective management of large-scale ecological systems. Loxahatchee/Hungryland Slough and the Everglades Agricultural Area watersheds directly north of the remaining Everglades system each receive a score of greater than 7.0 indicating a high degree of information sharing, joint database production, and other collaborative efforts among jurisdictions, organizations, and major landowners. The Lower East Coast and Manatee River watersheds (in the western portion of the state) also score high in terms of collaborative capabilities.

Again, watersheds to the west and southeast of Lake Okeechobee have the highest scores for the *Implementation* plan component. These areas contain some of the highest population figures and population growth rates for the state, as well as some of the most complex environmental problems. As a result, the public pressure to draft a strong environmental plan and also ensure that it is implemented may contribute to high implementation scores in these areas.

The Total Plan Quality (TPQ) score computed from the area-weighted sum of all plan components allowed us to evaluate the overall management capabilities of multiple jurisdictions for each watershed in the southern part of the state. This phase of analysis serves as a global assessment of the relative strength of ecosystem management capabilities from a spatially "bottom-up" perspective. Watersheds with above average total ecosystem plan quality scores generally occur in two clusters (Figure 7.3). The first concentration of high scores are located to the east and south of Lake Okeechobee, extending to the lower east coast of the state encompassing the urban corridor from West Palm Beach to Miami. The Loxahatchee/Hungryland Slough and the Everglades Agricultural Area watersheds are the highest scoring in the sample, indicating that these are prime areas to facilitate collaborative ecosystem management initiatives. High scores for these watersheds can be attributed to the strength of the Palm Beach County plan which is one of the highest scoring jurisdictions among selected comprehensive plans.

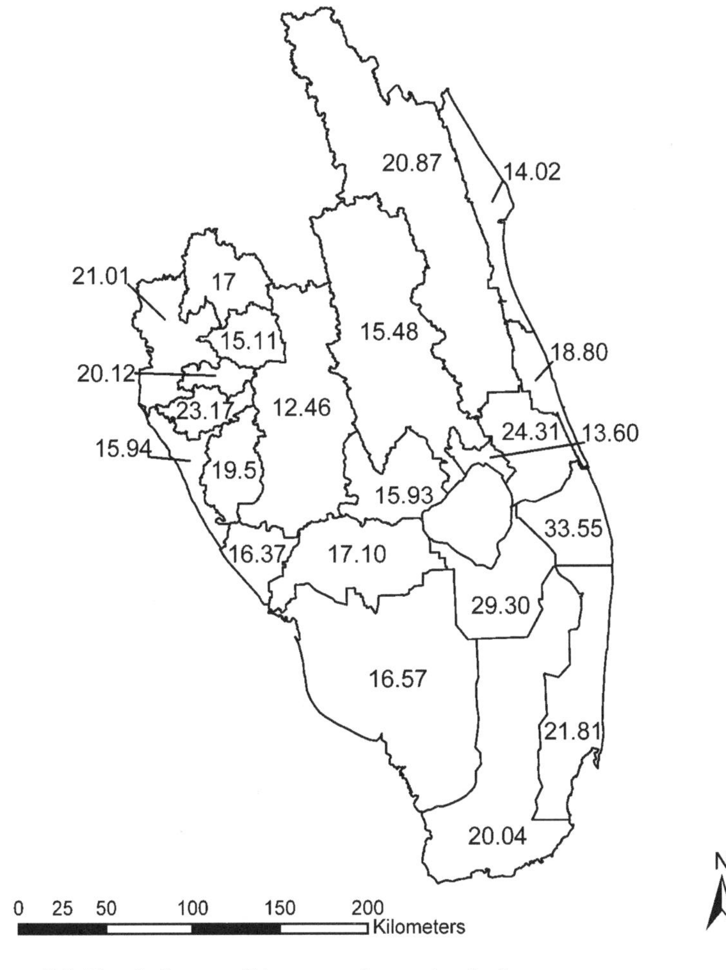

Figure 7.3 Total plan quality scores by watershed
Source: Brody et al., 2004: 45.

The second concentration of above average watershed scores occurs to the west of the study area in the greater Tampa Bay region. Tampa Bay, Manatee River, and Little Manatee Rivers watersheds all receive strong total plan quality scores. From a hydrological perspective these watersheds are associated with the Tampa Bay Estuary which, as mentioned above, is a natural system of national significance. In contrast, inland watersheds in the northern portion of the study area with lower levels of perceived biodiversity and lesser-known natural value receive some of the lowest total plan quality scores.

Factors Influencing Ecosystem Plan Quality Scores

Correlation analysis provides an initial step in understanding the major factors contributing to the ability of local jurisdictions to collectively manage ecological systems. Table 7.4 shows correlations between the measure of ecosystem plan quality and several environmental, socioeconomic, demographic, and other contextual variables.[4]

Not surprisingly, results for existing ecological conditions are similar to those found in the previous case. The proportion of area with high biodiversity within a watershed has no significant statistical bearing on plan quality. This result runs contrary to the assumption that areas of high biological importance would stimulate planners, some of the most proactive policy agents, to draft plans that seek to protect the integrity of these critical natural resources. However, human disturbance is a significant factor ($p<.05$) in raising total ecosystem plan quality scores. An increasing proportion of human disturbance within a watershed, such as pavement, agricultural practices, and the presence of invasive species leads to stronger watershed planning capabilities. Consistent with the previous case, only when biodiversity or critical habitat is under threat from anthropogenic stresses (e.g. urban development) does it appear to have a significant positive impact on plan quality. A notable exception is the *Factual Basis*. This plan component is not significantly correlated with measures of disturbance. Planners in highly urbanized local jurisdictions may not believe their natural resource base significant enough to inventory, analyze, and present in their plan.

Socioeconomic and demographic factors are also associated with measures of ecosystem plan quality. Wealth, as measured by the median home value within a watershed, is the most significantly correlated variable with total plan quality in the study ($p<.001$). Jurisdictions with wealthier populations usually have more financial resources to devote to planning staffs and plan development which leads to the adoption of higher quality plans. Furthermore, residents with high incomes often have more time and interest in participating in the planning process, particularly in regard to environmental issues (Van Liere and Dunlap, 1981; Scott and Willets, 1994; Fransson and Garling, 1999).

---

4 Due to the small sample size and lack of statistical power, we considered a correlation significant when $p<.1$. The small sample size also caused us to limit our analysis to correlations, which does not control for other factors.

Table 7.4 Correlations for total ecosystem plan quality

Variable	Correlation coefficient	P-value
Environmental variables		
Biodiversity	-.04	.85
Human disturbance	.40	.05
Socioeconomic and demographic variables		
Population	.44	.03
Population change	.39	.06
Wealth	.75	.00
Education	.72	.00
Land use variables		
Agricultural	.27	.19
Commercial	.06	.76
Industry	.30	.16
Multiple family	.16	.44
Single family	.05	.82
Contextual variables		
Planning agency capacity	.50	.01
No. of jurisdictions in watershed	.06	.78
Area of watershed	-.01	.94

Source: Adapted from Brody et al., 2004: 45.

Watersheds containing highly educated populations also help explain total ecosystem plan quality scores ($p<.01$). Past studies have linked levels of education and the degree of environmental concern (Howell and Laska, 1992; Guagano and Markee, 1995; Raudsepp, 2001). The results of this study may provide additional insights by suggesting that an educated public can influence the planning process, and encourage the adoption of plans that are focused on protecting the integrity of ecological systems over the long term. Population levels as of the year 2000 ($p<.05$) and population change between 1990 and 2000 ($p<.1$) are also positively correlated with total plan quality scores. These results are expected since high population levels are closely related to urban development and associated human disturbance. Growth pressures are associated with higher levels of disturbance to critical habitat, resulting in a greater perceived need to protect remaining areas of biodiversity.

Lastly, the proportion of five different types of land use within a watershed is not significantly correlated with plan quality scores where $p<.1$. This result may be attributed to the small sample size and increased difficulty in finding statistically significant correlations due to lack of statistical power. It could also mean that the type of land use is not as important as the specific impact on a given parcel. However, the degree of association for these different land uses varies. Agriculture ($p = .19$) and industrial ($p = .16$) land uses have a much stronger correlation with ecosystem plan quality than multi and single family residential. These results are consistent with the findings above since industrial and agriculture uses are closely related to human disturbance on ecological systems. A test of means between high intensity land use (commercial, agricultural, industry, and multi-family residential) and low intensity land use (single-family residential, estate, and preserve) is statistically significant where $p<.05$.

Perhaps the most useful finding for practicing planners is that greater planning agency capacity is significantly correlated with greater watershed planning capabilities as measured by ecosystem plan quality ($p<.05$). High numbers of planning staff are associated with increased levels of financial resources, expertise, and commitment to drafting a high quality environmental plan. This result is also consistent with the above findings that large, wealthy populations living in urbanized areas contribute to strong watershed planning scores. The finding may also indicate that small communities with understaffed planning departments are at a distinct disadvantage when it comes to protecting ecological systems from future development. A key recommendation (described in more detail in Chapter 12) stemming from this case analysis is to install an appropriate level of planning capacity via staff members before rapid growth and development takes place. This more proactive approach will ensure that the necessary amount of staff members and associated expertise are in place to allow regional development to take place in a more sustainable manner, or in a way that protects the integrity of ecological systems.

While the remainder of the contextual control variables is not significantly correlated with ecosystem plan quality scores, the results still provide insights on how to manage watersheds effectively at the local level. In this study, the number of jurisdictions in a watershed (ranging from 3 to 12) has no statistical bearing on the strength of plan quality scores. This result suggests that a relatively large number of jurisdictions within a single watershed may not compromise the ability of these jurisdictions to manage the entire ecological system collectively from the local level. An increasing number of parties will inevitably demand more collaboration and political will to accomplish transboundary management; however in this case study such hardships do not seem to adversely affect the degree to which the ecosystem is collectively managed. The size of the watershed also has no major statistical effect on ecosystem plan quality scores. One would expect larger ecological systems to be more difficult to manage at a local level (and in fact the correlation is negative), but we found no significant relationship between the areas of watersheds and plan quality measurements.[5]

---

5 It should be emphasized, however, that given the small sample size and lack of statistical controls, these results should be considered tentative. A larger study may yield different conclusions. Additional research and data are needed to fully understand the influence of these contextual control variables on watershed management capabilities.

Chapter 8

Collaborative Environmental Planning and Stakeholder Participation

Another major factor that will impact the quality of local plans to manage ecological systems is stakeholder participation. Because ecosystem approaches to management often extend across different organizations, agencies, and lines of ownership, the planning process usually necessitates the involvement of multiple and sometimes competing interests. Furthermore, many local comprehensive planning processes, such as those in Florida are by mandate required to develop a citizen participation program as part of their planning process. Who is involved and to what degree will inevitably influence the outcome of the process: the plan. Often times, the focus of collaborative environmental initiative is on intergovernmental relations, such as between various federal agencies or state and local government. But, coordination at the ecosystem level should incorporate the interests of the broader community to include non-government organizations, industry, private landholders, and local citizens. Without including all stakeholders in a framework of collaboration and joint problem solving, ecosystem management initiatives are bound to have limited success (Wondolleck and Yaffee, 2000).

High levels of public participation are often cited as a central component of an effective planning process for ecosystem management and environmental planning in general. Scholars argue that because ecosystem management is by definition a transboundary, multi-party issue, participation of key stakeholders is widely viewed as the single most important element of a successful outcome (Grumbine, 1994; Westley, 1995; Yafee et al., 1996; Duane, 1997; Duram and Brown, 1999; Mccool and Guthrie, 2001). Participation of stakeholders from the beginning of a project increases trust, understanding, and support for regional or ecosystem-based protection. Including key parties in the decision-making process also helps to build a sense of ownership over a proposal and ensures that all interests are reflected in the final management plan (Innes, 1996).

Public participation in plan making was initially practiced to reflect a commitment to the principles of democratic governance. As discussed by Arnstein (1969); Burke (1979); Day (1997); Fainstein and Fainstein (1985) and others, these principles support the rights of individuals to be informed, consulted, and to express their views on governmental decisions. They also include the need to better represent the interests of disadvantaged and powerless groups in governmental decision-making, and the contributions of participation to citizenship.

More recently, it has been argued that citizen participation can act as a powerful lever for generating trust, credibility, and commitment to the implementation of policies (Innes, 1996). Including key parties early, often and ongoing throughout the planning process helps build a sense of ownership over its content, reducing potential conflict over the long term because those involved become responsible for the policies set in place (Creighton, 1992). Furthermore, organizations and individuals often bring to the process valuable knowledge and innovative ideas about their community that can increase the quality of adopted plans (Moore, 1995; Beierle and Konisky, 2001). This "bottom-up" approach to planning, where information is shared widely among various parties, may not always lead to stronger protection of critical natural resources, but will help ensure that a proposal incorporates the values of those in the community, thereby reducing the need for costly enforcement measures later down the line (Crowfoot and Wondolleck, 1990).

Direct involvement, in this sense, moves beyond consultation to give interested parties responsibility for making decisions and taking credit for the implementation of policies (Arnstein, 1969). Kaza (1988) perhaps best summarized the strength of participatory planning when she stated, "with involvement comes understanding, with understanding comes public support and commitment." Godschalk et al. (1994) further notes that "while early involvement and participation adds time and cost at the initial planning stage, the up front investment often pays off when it comes to agreement and implementation. The results can be more fair, equitable and pragmatic; easier to implement; longer lasting; and less costly than alternative approaches" (p. 19).

For example, Yaffee et al. (1996) found that participation and collaboration of key stakeholders was the single most important factor (cited by 61 percent of respondents) that enabled projects to reach a quality outcome. Specifically, collaboration within and among public agencies and businesses was an important mechanism for increasing cooperation and communication, fostering trust, and allowing for a more effective outcome that met a greater set of interests. Furthermore, the most important suggestion offered by respondents (41 percent) was to involve all stakeholders in the planning process at an early stage. Participation of industry in the planning process was emphasized because it better enabled parties to share funding responsibilities and pool resources, reduce duplication of efforts, and promote more efficient use of limited resources.

In other research, Innes (1996) examined the role of consensus building through case studies of environmental problems involving multiple issues that crossed over jurisdictional boundaries. All the cases involved shared power across agencies and levels of government, and between private and public sectors. Innes found that not only did collaboration increase trust, communication, and the development of public-private networks, but that it resulted in stronger outcomes or plans that were beneficial to the resource or to the natural system as a whole. Furthermore, in a study of over 100 environmental disputes in which an agreement was reached, Bingham (1986) found that participation of key parties significantly influenced the likelihood that the agreement was implemented. More recently, Kennedy et al. (2000) discovered in their analysis of 100 cases involving watershed management in the US that collaboration by stakeholders was a key feature in improving resource management.

Based on this line of argumentation, a consensus-building planning process that seeks to generate ecosystem-based policies begins with the representation of key stakeholders (Carpenter and Kennedy, 1988; Crowfoot and Wondolleck, 1990; Beatley et al., 1994). Representation of a broad cross-section of the community includes industry and other private landowners, which in many instances are left out of important local land use decisions. In addition to the breadth of participants (a representative sample of the community) present in the planning process, active participation of specific stakeholders from the beginning of a project increases trust, understanding, and support for policies that protect natural systems and their sub-components (Duane, 1997; Yaffee and Wondolleck, 1997; Duram and Brown, 1999). Insofar as an adopted local plan represents the sum of collective interests, participation in the planning process must be considered when assessing the content and quality of the final product.

It should be noted that while a large portion of the literature strongly supports representation and participation of specific stakeholders in the planning process, few empirical studies exist to support these claims. And, there are counter arguments that suggest participatory process may not necessarily lead to a quality plan. High levels of participation may increase conflict by having disputing parties at the negotiating table, frustrate planners by slowing down the decision-making process, and most importantly dilute the strength of the final agreement by having to balance competing interests (Alterman et al., 1984; Brody, 2001).

The Influence of Participation on Ecosystem Plan Quality

Further explanation is needed as to why stakeholder participation may lead specifically to stronger local ecosystem plan quality. The participation of stakeholders is often associated with *land ownership*, *resources*, and *knowledge* that, when brought to the planning process, can increase the quality of the final plan. One of the fundamental goals of ecosystem management is to ensure that critical land within a natural system, such as a watershed, is included for management within the targeted planning area. When key landholders are active participants in a comprehensive planning process, areas of high biodiversity, natural habitat, or critical ecosystem components may receive greater consideration in the final plan. If large landholders are not part of the planning process, the final plan may not cover the entire ecosystem, falling short of its intentions to manage the natural system. Stakeholder participation can also contribute valuable resources, such as time, personnel, and sometimes funding, which will enhance plan quality by allowing for more expansive data collection, better monitoring programs, more regular plan updates, etc. Finally, with participation from a range of stakeholders comes knowledge of the resource and technical expertise that will inevitably contribute to higher plan quality. More than ever, private sector actors, such as industry are collecting and analyzing their own baseline data to monitor the natural resources upon which they depend.

The presence of certain stakeholders in the planning process can thus boost the collective capacity of planning participants, which should enhance each individual component of a plan. For example, it is expected that the factual base would include a more complete resource inventory, where impacts to these resources would be better known. Goals and objectives would be more inclusive, better balanced, and reflect a more system-wide approach. Inter-organizational coordination elements would be stronger where more collaboration with other parties and jurisdictions is emphasized. Tools and strategies would be more focused and inclusive, include more incentive-based policies, and better monitoring tools. Finally, implementation sections of the plan would provide greater accountability, flexibility, and enforcement of policies. The underlying assumption of the positive influence of stakeholder participation is that these groups have valuable knowledge and resources to contribute to plan development.

In summary, there are several major benefits identified by Gray (1989) and others that arise when participation and collaboration of key stakeholders take place to address problems that cut across public and private sector interests. First, collaboration increases the quality of the solutions, such as management plans considered by the parties because the collective capacity to respond to the problem is increased. In particular, private sector organizations contribute time, resources, and information to a collaborative situation which will strengthen the outcome of the planning process. Second, early participation in a multiparty collaboration can minimize the possibility of conflict that might occur during the later stages of a project. Third, the process of collaborating builds in certain guarantees that each party's interests will be protected. Participation will increase individual leverage over a problem and ensure that the outcome meets the objectives of those involved. Fourth, collaboration builds a sense of ownership over the problem that will raise the level of commitment to implementing the solution. If an industry or another stakeholder is part of a planning process it will more likely carry out the solution contained in a management plan or agreement. Finally, participation facilitates a two-way education process between planners and community members that may increase the planning outcome (Howell et al., 1987). In Pinellas County, Florida, for example, government staff was able to learn more about the habitats they were seeking to conserve while at the same time they were able to educate participants about the specific goals of environmental planning. Educating each other helped build collaborative relationships that in turn helped develop a strong environmental component to the comprehensive plan.

Resource-based Industries in Ecosystem Approaches to Management

Most of the literature on stakeholder participation in planning and ecosystem management is written primarily from a public sector perspective where the influence of government or non-government organizations is examined. The participation of industry does not receive a great deal of attention in arguments for collaboration and consensus building, despite the fact that industry has the largest impact on our natural resource base and that much of the critical habitat in the US is located on industry-owned lands (Wondolleck and Yaffee, 2000; Brody, 2003).

Industry land holdings (a subset of privately-held lands in the US) include many important elements of ecosystem diversity, particularly in the eastern part of the country, and comprise approximately two-thirds of the land base of the continental US. So, government must encourage industry participation to adequately protect biodiversity (O'Connell, 1996; Vogt, et al., 1997). For example, 57 percent of forests in US are privately owned. In regions such as the southeast, private ownership comprises up to 90 percent of the land base. Furthermore, 90 percent of the more than 1200 listed endangered and threatened species occur on nonfederal lands and more than 5 percent, including nearly 200 animal species, have at least 81 percent of their habitat on nonfederal lands (Wondolleck and Yaffee, 2000).

Consistent with these data, Cortner and Moote (1994) argue that a fundamental requirement for effective ecosystem management is the coordination of public and private interests. Hoffman et al. (1997) suggest that because much of the critical habitat in the US lies on business-owned land, the inclusion of this key stakeholder in the decision-making process is necessary to achieve successful management of ecological systems. They further assert that involving business- related stakeholders is the best way to foster joint gains in environmental protection and economic growth over the long term.

These arguments are supported by data from Beyer et al. (1997), who found that the informal participation of industrial forest stakeholders was one of the keys to the present and future success of the Eastern Upper Peninsula of Michigan Ecosystem Management Project (EUP). This group is comprised of government agencies, forest product companies, and the Nature Conservancy, a leading environmental Non-Governmental Organization (NGO). These partners (composed of eight public and private landholders) collectively manage 2.6 million acres of land in the EUP. Despite varying resource management goals and activities, group members have formed a collaborative venture to facilitate the sustainable management of the EUP ecosystem over the long term. In summary, private lands clearly must play a critical role in any cooperative strategy to protect biodiversity and ecosystem integrity.

Jones (1994) notes that the concept of ecosystem management is virtually untested within the ownership pattern that dominates much of the eastern US: non-industrial private forests (NIPF). He argues that if ecosystem management is to ever become more than a theoretical construct, more studies must be done on how to include NIPF owners in the decision-making process, encourage them to look beyond their property lines, talk with their neighbors, and consider collaborative approaches to managing critical ecosystem elements. Whether it involves an industrial operation or a large land holding developer, there is an increasing emphasis on private sector participation in planning for critical natural resources. Several large corporations, such as Mead, Champion International, and Westvaco are slowly realizing the benefits of collaborative ecosystem management, and are devoting time and resources to developing ecosystem-based projects on their land.

The importance of the large landholding industry in ecosystem management is also outlined by Machlis (1999) who states that for ecosystem management to be successful, key institutions such as resource management agencies, governments, and corporations with large land holdings should be included in the planning process. Machlis also notes the timber industry's increasing interest in achieving best management practices for biodiversity while pursuing other management objectives such as timber production. MacKenzie (1996) further discusses the importance of industry as a stakeholder in her study of ecosystem approaches for restoring the Great Lakes ecosystem. She stresses the significance of industry in the achievement of planning goals and in the remediation process of the lake ecosystem. Stakeholders in several of her studies felt the inclusion of locally owned industries in the ecosystem planning process may yield positive results by increasing community identification and ownership of the plan.

Chapter 9

Measuring the Effects of Stakeholder Participation on the Quality of Local Plans

The previous chapter argued that public participation and involvement during the planning process is an essential component of effective ecosystem management. Since ecosystem approaches to management follow ecological boundaries, rather than administrative or political lines, stakeholder input, collaboration, and the formation of partnerships across land ownership are an essential part of reaching a desirable outcome. While theorists and practitioners consistently call for widespread participation in ecosystem management and environmental planning in general, few, if any studies have empirically tested the assumption that representation and participation of stakeholders during the planning process will lead to a stronger management plan.

In response, this chapter presents an in-depth case examining the influence of stakeholder participation in ecosystem approaches to management. Specifically, it quantitatively tests the effects of stakeholder representation and participation on our ecosystem plan quality measure in Florida. In addition to the overall breadth of stakeholder groups involved in planning, the effects of specific stakeholders are tested and discussed to determine which has the greatest impact on the quality of the adopted plan. Examining the statistical impact of participation during the planning process on the quality of plans can not only support or contradict the theoretical arguments and case study analyses pervading the literature, but may also add insight into how plans can be strengthened by considering who specifically is involved in the planning process. Better understanding the relationship between the planning process and planning outcome will enable communities to more effectively manage their ecological systems and critical natural resources in the future in Florida and in other states with public participation requirements.[1]

To investigate whether the representation and participation of stakeholders do in fact strengthen the quality of the planning outcome as applied to ecosystem approaches to management, we can posit the following hypotheses emerging from Chapter 8: 1) Representation of key stakeholders in the planning process will result in a higher quality plan; and 2) Participation of specific stakeholders, such as industry,

---

1 Florida, Georgia, Hawaii, Maine, Maryland, New Jersey, Oregon, Rhode Island, Vermont, and Washington all have widely recognized state growth management programs that either require or strongly encourage the adoption of local comprehensive plans (Brody et al., 2003).

government, and NGOS will result in a higher quality plan. The first hypothesis tests the general assumption that stakeholder representation (breadth) leads to a stronger plan. The second is more specific in that it focuses on the effects of stakeholders (activity) participating in the planning process.

Once again, the same random sample of 30 local jurisdictions generated for Case 1 in Chapters 5 and 7 was used for this case to isolate the influence of stakeholder presence on ecosystem plan quality. In addition, the same five component plan coding protocol and measurement procedure was used to derive an overall score for ecosystem plan quality based on a scale of 0-50 (for more detail see Case 1, Chapter 5).

Measuring Stakeholder Groups

Stakeholder participation variables were measured through a survey on public participation and planning conducted as part of a National Science Foundation (NSF) research project. In each jurisdiction, personal interviews with planning directors and citizen participation staff were conducted to measure characteristics of the participation processes. Information was obtained on the level, timing and extent of thirteen different stakeholder groups, ranging from environmental non-government organizations (NGOs) to local neighborhood groups.[2] The representation variable was measured as the percentage or breadth of these stakeholders present during the planning process (total number of groups present in the process divided the total number of groups recorded).

The participation variable was created by grouping a subset of the thirteen stakeholders into the following five core participant categories: resource-based industry (agriculture, forestry, marine, etc.), business (i.e. development associations, commercial development groups, homeowners associations), environmental non-government organizations (NGOs), local government, and others (e.g. neighborhood groups, elected officials, affordable housing groups, representatives of special districts, etc.). The construction of these categorical variables enabled us to examine the effects of the active participation of specific groups, rather than simply an overall measure of representation.

The Benefits of Participation

The analysis of stakeholder participation and ecosystem plan quality was conducted through three lenses of focus: broad representation of a large number of stakeholders, targeted participation focusing on five stakeholder groups, and the addition of participation contextual factors to control for alternative explanations of the variation in plan quality. With each increasing level of focus or specificity, the impacts of stakeholder participation in the planning process become better understood and the

2 The presence of these 13 different stakeholders in the planning process was recorded as a dichotomous or "dummy" variable.

conditions of when participation is most effective in producing high quality plans becomes clearer.[3]

Despite a strong theoretical justification broad representation of stakeholders, ranging from the agricultural industry to neighborhood groups, does not have statistically significant influence on plan quality (Table 9.1). Simply having a wide range of participants present in the planning process does not seem to guarantee higher quality plans when it comes to managing ecological systems. One explanation for this finding is that competing interests and a planning process burdened by multiple groups wanting to voice their opinions may hinder the quality of the outcome. Broad and diverse stakeholder participation can thus lead to a "lowest common denominator" when it comes to plan quality because there are fewer opportunities for agreement. For example, in Sarasota, the decision-making process was shackled by the multitude of participating stakeholders because elected officials were so open to citizen concerns and allowed for such lengthy discourse over pertinent issues. Allowing every vocal interest to speak or comment slowed down the planning process, frustrated many participants, and at times diminished the ability of both the Planning Board and the City Commission to make quick decisions (Brody, 2001a).

Table 9.1 Representation in the planning process

Variable	Coefficient	Standardized coefficient	Standard error	T-value	P-value
Representation	7.75	.25	5.77	1.343	.190
Constant	17.21		2.90	5.931	.000
N:	30				
F-Ratio (1,28):	1.80				
Significance:	.1899				
R-squared:	.0605				

Source: Adapted from Brody, 2003: 6.

While broad representation does not have a significant impact, the presence of individual stakeholders does statistically impact the quality of comprehensive plans with regard to their ability to protect natural systems (Table 9.1). The presence of resource-based industry groups (agriculture, forestry, marine, and utilities) has the strongest positive influence on ecosystem plan quality. As shown in Table 9.2, with

3 Several statistical tests for reliability were conducted to ensure the OLS estimators were Best Linear Unbiased Estimates (BLUE). Tests for model specification, multicollinearity, and heteroskedasticity revealed no violation of regression assumptions. In addition, a series of diagnostics were performed to test for influential data points or outliers in the data set. Given the small sample size, influential data points may have a significant impact on the interpretation of ecosystem plan quality. Various types of plots, as well as robust regression uncovered no influential data points affecting the results.

each added industry group in the planning process, on average, the final planning score will jump 10 points (which is statistically significant compared to the baseline variable others at the .05 level). While industry participation is rare, the occurrence provides a planning boost almost twice more powerful than an environmental NGO.

This finding supports the idea presented in Chapter 8 that although resource-based industry is often overlooked as a key stakeholder, it brings to the planning process valuable knowledge and resources regarding its ownership of critical habitats, which in turn increase the quality of adopted plans. This notion is important because as already stated, industry not only has the largest impact on our natural resource base, but also much of the critical habitat in the US is located on private lands. Because public lands do not include many important elements of ecosystem diversity, particularly in the eastern part of the US, and comprise only one-third of the land base of the continental US, protecting biodiversity at all levels of government will rely on industry participation.

Case study research of planning processes based on site visits to several jurisdictions in Florida supports the statistical findings. For example, the participation of the marina industry in the Fort Lauderdale planning process resulted in stronger coastal management policies. Marine trade and recreation representatives met in groups and one-on-one with planning staff throughout the development of the comprehensive plan. Since this stakeholder group depends on a healthy natural environment for its growing business, it has a financial interest in ensuring clean waters. The marine industry proposed higher water quality standards and clean up efforts that were incorporated as policies in the final plan (Brody, 2001b). In this instance, industry was a driving force in generating stronger environmental and ecosystem management policies for coastal areas.

Similarly, in Pinellas County, the Florida Power and Light Company (FPL) played a key role in educating planners about existing natural resources and generating policies to manage those resources for the future. As a major landholder and community member, FPL was an active participant in the planning process. The company shared information related to critical habitats on their lands and ensured that these areas were considered part of the environmental programs associated with the plan. More specifically, FPL allowed critical habitats occurring along utility easements to be incorporated into the existing network of protected lands throughout the county (Brody, 2001c).

It is important to note that of course not all resource intensive industries make environmental protection a priority. In fact, organizations in Florida and across the country have violated environmental regulations and possibly concealed the impacts of their operations on the natural environment to avoid costly lawsuits. Large single firms can also have diverse and conflicting interests and behaviors so that they can simultaneously violate environmental regulations while becoming involved in initiatives to protect critical natural resources on their lands and the lands of others. However, there is evidence to support the hypothesis that under specific circumstances, when industrial groups want to be a part of the planning process, their participation can significantly increase the environmental quality of the adopted plan.

The presence of non-government organizations (NGOs) in the planning process also has a significant positive impact on plan quality ($p<.05$) compared to the baseline dummy variable. This result is expected, since environmental groups often provide valuable environmental data and expertise to the planning process. The pro-environmental stance and educational mission of many NGOs should drive ecosystem plan quality higher. For example, by actively participating in the Pinellas County planning process through a working group, the Audobon Society was able to educate County staff by sharing their data and environmental knowledge of the region. In this case, communication, information sharing, and a staff receptive to the comments of working group members led to a stronger, more innovative environmental component of the comprehensive plan. By initiating a two-way exchange of ideas, all parties were able to more effectively meet their environmental management goals and produce a balanced plan reflecting a diversity of interests. Through environmental working group discussions, it was pointed out by the Audobon Society that existing parks served as migratory bird habitat (Brody, 2001c). Certain activities by park staff, such as mowing native vegetation, were detrimental to the bird populations. These concerns led directly to a policy in the final plan (*policy 3.1.6*) that strengthens the level of protection for critical habitats in existing parklands.

Surprisingly, the presence of local government departments in the planning process has a negative effect on ecosystem plan quality. Although the effect is not statistically significant, it would be expected that the participation of government agencies would increase the quality of the plan. One explanation for this result is that aside from environmental departments, government agencies such as transportation or public services tend not to have the long-term management of the natural environment as a prime interest. Furthermore, the participation of multiple government departments could delude the strength of the final plan through competing interests or conflicting planning goals.

Table 9.2 Key stakeholders in the planning process

Variable	Coefficient	Standardized coefficient	Standard error	T-value	P-value
Industry	10.06	.58	2.60	3.862	.001
Business	3.54	.18	2.60	1.366	.184
NGOs	5.06	.33	2.34	2.166	.040
Government	-3.05	-.177	2.58	-1.185	.247
Constant	13.16		2.81	5.391	.000
N:	30				
F-Ratio (4,25):	7.77				
Significance:	.0003				
Adjusted R-squared:	.4829				

Source: Adapted from Brody, 2003: 7.

Overall, examining the effects of key stakeholders taking part in the planning process, rather than broad representation is a more effective approach to understanding how participation influences ecosystem plan quality. The land, knowledge, and resources specific groups bring to the planning process can greatly increase the quality of plans. When specific stakeholder groups whose interests are aligned with the plan evaluation criteria participate in the planning process, ecosystem plan quality will improve. The challenge to planners, then, is to identify and target for involvement in the planning process which groups will increase the quality and performance of the adopted plan.

Next, contextual control factors were analyzed along with the most significant stakeholders to further isolate the effects of industry participation on ecosystem plan quality. Wealth, population, and planning capacity (i.e. the number of staff devoted to drafting the comprehensive plan) were included to control for extraneous variables that may also drive the plan quality measure. As shown in Table 9.3, resource-based industry participation remains a powerful predictor of ecosystem plan quality. The population of each jurisdiction is the most significant variable in the analysis, which may be explained by the fact that population levels can often be associated with increased urban development and decline of critical habitats or overall biodiversity. As shown in Chapter 7, growth pressures associated with higher levels of disturbance to natural ecosystems, resulting in a greater perceived need to protect remaining areas of biodiversity. High levels of population may in this case indirectly drive ecosystem plan quality higher.

Table 9.3 Key stakeholders and contextual controls in the planning process

Variable	Coefficient	Standardized coefficient	Standard error	T-value	P-value
Industry	6.82	.40	2.45	2.784	.010
NGOs	5.00	.03	2.13	.235	.816
Wealth	-.57	-.011	6.27	-.091	.928
Population	8.05	.64	1.98	4.049	.000
Capacity	-.42	-.17	.32	-1.326	.206
Constant	-13.72		34.76	-.395	.000
N:	30				
F-Ratio (4,25):	12.77				
Significance:	.0000				
Adjusted R-squared:	.7269				

Source: Adapted from Brody, 2003: 8.

Interestingly, the significant effect of environmental NGOs on ecosystem plan quality is lost with the addition of contextual controls. This result may be explained by the high correlation between population and the presence of environmental NGOs. On average, most large environmental groups with the ability to boost the collective capacity of the planning process are located in urban areas or jurisdictions with large populations. Thus, with the inclusion of population levels in the model, the positive impact of NGOs on ecosystem plan quality is washed out. This result could also reflect stronger environmental values typically present in urban populations that can support the presence of environmental NGOs.

Putting Participation into Perspective

Although the representation of stakeholders during the planning process may play a role in increasing the likelihood of plan implementation, based on this case it is not a significant factor when it comes to producing a high quality outcome within the context of ecosystem or environmental planning. Despite the broad theoretical support for representation as a basis for sound planning, the evidence suggests that having all of the stakeholders and community members present during the decision-making process does not necessarily guarantee the adoption of a strong plan (Brody, 2003). For practicing planners, then, there is an apparent dichotomy between linking the planning process to outcomes or to plan implementation. If environmental planners are interested in generating the highest quality plans to manage ecological systems over the long term, then broad stakeholder representation is not necessarily beneficial and, in some cases, can be detrimental to plan quality. It may be that planners could have to make a choice between generating high quality environmental plans or generating plans that will be supported and implemented in the future. Having all community interests "on board" may increase the chances that the final plan will be implemented, but it may end up being a weak regulatory document. That being said, planners should not concentrate on involving fewer stakeholders during the planning process. However, instead of being concerned about the number of stakeholders involved and ensuring there is complete representation of the public, planners may instead want to focus on incorporating specific groups that will most likely boost the quality of the adopted plan.

While broad representation of stakeholders in the planning process does not necessarily lead to stronger plans, despite the endorsement of many scholars (Crowfoot and Wondolleck, 1990; Beatley et al., 1994; Beierle, 1998; Susskind et al., 1999), the presence of specific stakeholders does in fact significantly increase ecosystem plan quality. While environmental NGOs are expected to raise plan quality since their goals are often to protect ecosystems, a significantly positive impact from resource resource-based industry participation is somewhat surprising considering its historical battles against environmental protection initiatives, particularly in Florida.

This finding is critical because it demonstrates that when engaged in the planning process, resource-based industry has an interest in environmental management and brings to the negotiating table valuable knowledge and resources, which ultimately lead to a stronger comprehensive plan. Increasingly, large resource industries, such as forestry and agriculture are becoming involved in environmental planning processes because: 1) they realize that maintaining the economic viability of their operations relies on sustainably managing and even protecting their natural resource base; 2) demonstrating environmental concern can result in favorable media attention and public support for their business activities; and 3) participating in a collaborative process can facilitate information and data sharing that will in turn improve the performance of commercial operations. Resource-based industries may even provide the need boost in adopting a local plan that considers the broader ecological system. That is not to say all industries are concerned with ecosystem management and will help raise the quality of plans through active participation. Many organizations in Florida and around the country are staunchly opposed to any type of environmental initiatives since they view them as threats to corporate profitability. However, it is clear from the results of this case study that when industry groups want to be part of the planning process, they tend to positively impact the quality of the final plan as it relates to managing ecological systems.

PART 3
Plan Implementation

Part 3 of this book focuses on one of the most critical yet under-studied aspects of environmental planning and management: plan implementation. Identifying the components of a high quality plan that seeks to protect the integrity of natural systems over the long term is a futile endeavor unless the adopted plan is implemented. Chapter 10 describes the importance of implementation in environmental planning and the lack of attention it receives from both practitioners and the planning literature. Specifically, the debate over plan performance and plan conformance is highlighted. Chapter 11 presents two case studies that map and measure the degree of local plan implementation in Florida within large watersheds over a ten year period. The methods and results from this case provide a spatial compass for local environmental planners to ensure development patterns adhere to the original design of the adopted plan.

Chapter 10

Evaluating Plan Implementation and the Controversy over "Conformity"

Planning scholars and practitioners have long debated the importance of tracking and measuring the implementation of adopted policies. While evaluation and implementation techniques are well developed in the policy sciences, understanding how to assess implementation remains an elusive endeavor in planning and is often criticized as a major shortcoming in the field. Lack of data, methods, and empirical enquiry makes it difficult to respond to critics who consider plans to be "dead on arrival" or paper shells that are never put into action (Clawson, 1971; Calkins, 1979; Bryson, 1991; Talen, 1996a; Burby, 2003). Carefully crafting a planning process leading to an adopted plan that sustainably manages ecological systems over the long term is meaningless if the plan is not put into effect. How can planners validate the importance of plan making if they cannot determine if their plans have an impact on the community after they are adopted?

Up to this point, the book has been focused on measuring and predicting plan quality as an indicator of implementation; now we turn our attention to the quality of plan implementation itself. It is increasingly recognized that the strength of adopted plans does not necessarily correlate with implementation of their contents and that research is needed to understand the degree to which policies are being implemented after plan adoption. However, to raise the profile of environmental planning as a legitimate, if not essential policymaking endeavor, techniques must be developed to rigorously measure the efficacy of plans and degree of local plan implementation over time. Until we can evaluate the influence of plans subsequent to adoption, planning will remain an uncertain science.

The Need for Benchmarks

As often noted by Talen (1996a; 1996b; 1997), the fields of policy analysis and program evaluation have developed specific methods and a large research base on implementation, but there is a lack of parallel inquiry on implementation processes in the planning domain. This relative scarcity of research is particularly evident for plans that serve as blueprints or guides for the future physical development of urban areas. In these cases, there is little understanding of the relationship between the processes of planning, the adopted plan, and plan implementation or performance (Alterman and Hill, 1978). As a result, the field of planning seems to this day to be mired in what Calkins (1979) referred to as the "new plan syndrome," where plans and policies are adopted without any attempt to measure the progress toward achieving stated goals and objectives. Furthermore, no effort is made to determine why a previously adopted plan is unable to meet its goals even if they are partially or totally met.

The lack of systematic evaluation of plan implementation may be a consequence of several major obstacles facing planning scholars. First, it is unclear exactly when the outcome of a plan should be determined and what this outcome should be compared to (Baer, 1997). Since plans tend to be long-term policy instruments, it is difficult to establish a time frame for evaluating success. Restored wetlands, for example, could take decades to be considered ecologically functional. Furthermore, since the value of planning may be measured by more than plan content alone (e.g. planning process, social interaction, learning, etc.), there is disagreement over how to derive a measure of planning effectiveness. Second, the lack of longitudinal datasets, explicit research methods, and impatience of researchers to examine planning impacts over large time frames can also account for the lack of studies on the implementation of plans. Baseline data from which to detect change and measurable performance indicators are needed before systematic evaluation of community planning can occur (Murtagh, 1998; Seasons, 2003). Talen (1996a) argues that "methodological complexities alone are enough to thwart any evaluative endeavor" (p. 249). She finds a particular scarcity in the number of quantitative assessments of implementation success in planning.

A third obstacle is the debate over the meaning of planning success and the evaluation of plan conformity. Conformity measures the degree to which decisions, outcomes, or impacts adhere to the objectives, instructions, or intent expressed in a policy or plan. Alexander and Faludi (1989) reject this means-ends approach to measuring plan effectiveness because, due to the complexities of the decision-making process, deviation from a plan's original design is a normal consequence of policy implementation. Additionally, policy statements are meant to undergo modification in response to uncertain political and socioeconomic conditions. Under these arguments, the mere consultation of a plan may be viewed as an indicator of implementation success. Mastrop and Faludi (1997) reinforce this stance when discussing the merits of evaluating strategic plans. The authors assert that the established policy or plan should never be followed blindly but rather needs to be constantly reenacted and readjusted. Instead, the key to plan performance is the way in which a strategic plan holds its own during the deliberations following plan adoption. As previously noted, ecosystem-based plans should be regularly revised and updated under the principles of adaptive management, making them flexible policy instruments that must adjust with changing external conditions.

It is important to note, however, Faludi (2000) later distinguishes between strategic plans and project plans. While strategic plans are open and flexible, a project plan is a "blueprint" for the intended end-state of physical development. Once adopted, these plans are meant to be unambiguous guides to action where outcomes must conform to the specifications detailed in the plan. Faludi (2000) further elaborates that the evaluation of a project plan must follow the logic of ends and means and conformance of outcomes to intentions. Driessen (1997) supports this argument by concluding the criterion of conformity is unsuitable for assessing the performance of spatial planning policies *unless* the plan explicitly states (and all planning participants concur) that outcomes should conform to the original policy proposal.

At the other end of the plan implementation spectrum is the belief that plan intent and policy outcomes should follow a strict linear association (Wildavasky, 1973). Any departure from the goals and objectives of the adopted plan would, under this line of thinking, be considered a failure. Due to the uncertainties involved in the planning process, and the social and political complexities of plan implementation, a direct cause and effect relationship may be an unrealistic expectation for most plans. The real value of environmental plan evaluation can most likely be found not at the extremes, but somewhere towards the middle of the implementation spectrum. That being said, not holding planners and planning participants accountable for their adopted policies would be to undermine or de-legitimize the field of planning. Talen (1996a) asserts that the dismissal of linear association between the adopted plan and its outcome on the basis of uncertainty "can be seen as evaluation avoidance" (p. 254). In this sense, ecosystem planning policies must be adaptable to changing circumstances, but their intent must also be implemented and enforced in the field.

While the difficulties involved in evaluating plan implementation have restricted the focus of most empirical planning studies to measuring plan quality (see Berke and French, 1994; Burby et al., 1997; Burby and May, 1998; Brody, 2003a, 2003b, among others), there have been some past attempts to specifically measure the degree of plan implementation. For example, in Israel, Alterman and Hill (1978) conducted perhaps the most comprehensive study on plan implementation by measuring the degree to which plans conform to their original spatial design. Using building permits as an indicator of plan implementation, the authors found the level of accordance with the master plan in their study area was approximately 66 percent of the land area planned. They also used statistical models to explain the variation in plan conformity through several variables such as time and flexibility. Calkins (1979) presented a "planning monitor" to measure the extent to which plan goals and objectives are met, to explain the differences between plan and actual states of the environment, and to understand the reasons for any observed differences between the plan and the outcome. Using algebraic expressions, Calkins not only showed how to evaluate the overall plan, but also whether the desired spatial distribution has been achieved. This was the first attempt not only to measure if policy implementation conforms to the adopted plan, but also to identify where any discordance may occur. Such an approach is particularly relevant when evaluating plans that guide the physical development of a community, such as a local comprehensive plan.

Talen (1996b) builds on Calkins work by employing GIS and spatial statistical analysis to compare the distribution of public facilities called for in a plan with the actual distribution that occurred after plan implementation in Pueblo, Colorado. Mapping relationships between access to facilities as denoted in the plan and actual access years later revealed areas of the city that did not match the policymakers' original intent. More recently, Burby (2003) examined 60 local jurisdictions in Florida and Washington to explain the relationship between stakeholder participation in the planning process and implementation of natural hazards policies. By studying the ratio of proposed hazard mitigation actions that were subsequently implemented to proposed actions that were not implemented, Burby found that greater involvement of stakeholders in the planning process significantly improved implementation success. Lastly, Berke et al. (2006) tested the implementation debate using the development permit review process for stormwater mitigation in New Zealand. The research team used two conceptions of success in plan implementation: conformance and performance. The conformance approach defined success in terms of the degree to which permit review decisions precisely conform to the specific policies of plans as posed by Alexander and Faludi. The performance approach maintains that the prime concern is not whether there is a direct link between specific plan policies and implementation decisions, but rather how decisions help the community progress toward outcomes that resolve planning issues. When implementation was defined and measured in terms of conformance, plans and planners had an important influence on implementation success. Alternatively, if implementation was defined and measured in terms of performance, plans and planners were less influential in implementation. Clearly, assessing plan implementation will rest upon the definition or expectation of implementation. However, going through the trouble of developing a plan through a community-based process so it can sit on the shelves collecting dust while its merits are debated seems like a fruitless endeavor.

Major Factors Contributing to Nonconforming Development

Increasing the chances of plan implementation may not rest simply on the way the concept is defined, but rather on understanding and then activating the factors contributing to implementation success. Alterman and Hill (1978) and Burby (2003) both modeled the effectiveness of implementation using contextual variables such as population and population growth. Berke et al. (2006) add other planning-oriented variables to their statistical model, including planning capacity, enforcement style, and perhaps most importantly plan quality. The hope for planning scholars and practicing planners alike is that high quality plans are more likely to be implemented. Fortunately, the Berke et al. study found in their particular New Zealand context, when it comes to conformity, the quality of the plan in place leads to a significantly greater degree of implementation.

We can also draw from the growing literature on spatial development patterns and the influences of sprawl to help construct a statistical model for nonconforming development. In many instances, nonconforming growth patterns are manifested as urban and suburban sprawl. As growth spirals outward from existing urban centers, development infringes upon rural or protected areas or takes place in locations not intended by the jurisdiction's land use plan (Brody and Highfield, 2005). Since it appears that the same factors driving outwardly expanding growth patterns also contribute to the formation of nonconforming development clusters, we can gain insight from this literature to better understand what may be influencing the implementation success of ecosystem-based plans.

For example, Pendall (1999) acknowledges that land value is one of the most significant drivers of development and that sprawl occurs where land values are lower. Administering a survey in 25 metropolitan areas over 180 counties, Pendall found that high housing prices led to more compact development. Given that high housing values both reflect and perpetuate high land values, higher densities result with increased land values. Brueckner (2000) also cites the importance of land value in the urban expansion of cities. She states that "land conversion is guided by the economist's 'invisible hand' which directs resources to their highest and best use" (p. 162). Therefore, agricultural land will be preserved only if its productive value is worth more than the developer is willing to bid.

Economists identify three underlying forces that interact with land values to create spatial urban expansion or sprawl. First, population growth results in the outward expansion of urban areas. Second, rising incomes allow residents to purchase greater living space. These residents locate where housing options are less expensive, such as in suburban and ex-urban areas generally located at the periphery of metropolitan areas. Third, decreasing commuting costs produced by investments in transportation infrastructure also fuels outward expansion of development.

Socioeconomic and demographic characteristics are also considered important contributors to sprawling patterns of growth. For example, Carruthers and Ulfarsson (2002) show that population density influences the spatial extent of developed land. Development from a regional perspective becomes more compact as the number of people and jobs per acre increases. Increasing wealth further exacerbates urban expansion by allowing residents to purchase larger houses and properties (see also Alonso 1964; Brueckner 2000; Heimlich and Anderson 2001). With a high demand for low density, single-family housing developments, residents seek to locate where housing options are inexpensive, such as in the suburbs along the urban fringe. Daniels (1999) concurs stating that the "rising affluence of many Americans really drives the development of the fringe, because as income increases, the choices of what to spend money on expands as well" (p. 40).

Carruthers and Ulfarsson (2002) evaluated 283 metropolitan counties in the US at three points in time to examine the relationship between government fragmentation and several measurable outcomes of urban development, including per capita income. The study showed that income works to lower densities, spread out development, increase the amount of urbanized land, and increase property values. In contrast, Carruthers (2003) evaluated 822 metropolitan counties in the continental US between 1992 and 1996. Results from this analysis indicated that per capita income is only occasionally significant for determining the amount of growth at the urban fringe.

In addition to population density and rising incomes, race has been identified as another socioeconomic indicator of urban and suburban sprawl. Racial strife in the centers of cities such as Los Angeles and Detroit led to an out-migration of middle and upper class whites to the urban fringe (Daniels, 1999). This relocation of residents soon became known as "white-flight." Pendall (1999) analyzed this "white-flight" hypothesis and found that low-density zoning led to a decrease in construction of attached and rental housing. This trend in turn caused rents to rise, leading to a decrease in the population of Hispanics and Blacks in less compact development areas. Carruthers and Ulfarsson (2002) support the "white-flight" hypothesis showing that it is marginally associated with greater overall densities, more spread out metropolitan areas, and lower property values. Carruthers (2003) provided similar conclusions that race has a substantial effect on the spatial pattern of urban development.

Finally, several researchers have considered age as a factor in determining the spatial pattern of development. Specifically, Zhang (2001) found that younger residents are significantly related to new housing development. While other studies have shown that age is an insignificant predictor, the direction of the coefficients are consistent with the expectation that younger families promote sprawl and nonconforming development patterns by seeking out affordable housing options at the urban fringe.

In addition to socioeconomic factors, decreases in commuting costs due to infrastructure investment are another underlying force in the sprawling expansion of cities (Brueckner, 2000). Alonso (1964) cites improvements in transportation infrastructure as one of the primary reasons for a city expanding outwards. Daniels (1999) supports this idea, noting that new road construction will provide more access to the fringe. Heimlich and Anderson (2001) state that infrastructure drives the growth of cities by providing the essential framework for development. Once new development takes place, residents then demand improvements in infrastructure, which further ignites development along the urban fringe. Widespread access provided by improvements in transportation infrastructure allows developers to utilize cheap land located outside the city center (Gillham, 2002). Carruthers and Ulfarsson (2002) and Carruthers (2002) also found that per capita spending on road and sewer systems influence the spatial extent of development.

In contrast, survey findings by Pendall in a 1999 study of 25 metropolitan areas over 180 counties showed that investments in infrastructure, particularly heavy highway spending, did not lead to less compact development. In a study published in 2003, Carruthers found that infrastructure investments had mixed effects on growth at the urban fringe. Roadway investments appeared to have no impact on growth in suburban counties, while per capita spending on sewerage products occasionally led to greater growth at the urban fringe.

Lastly, land use planning and growth management policies have been theorized as determinants of the spatial pattern of development (Bengsten et al., 2004). Local policies, such as clustering of development, conservation easements, transfer of development rights, and urban growth boundaries have been suggested as strategies to reduce sprawl and promote a more compact form of development (Pendall, 1999; Mattson, 2003). These policies are likely to help guide growth in an ecologically sustainable manner and assist local communities in attaining the intended spatial design and land use intensities designated in their plans. The absence of such policies may allow for more sprawling development patterns involving an increasing loss of wetlands and leading to a greater degree of nonconformity. However, empirical studies are mixed. Shen (1996) found that growth management controls actually promoted sprawling development in outlying parts of Solano County, CA. In contrast, Knaap (1985) showed that the use of urban growth boundaries in Oregon contributed to increased density in urban areas and facilitated conforming development patterns. In most cases, all researchers note that a single growth management policy is not enough to mitigate outwardly expanding development, but must be installed as part of a broader program.

This chapter has argued that a better understanding of how to measure the degree of plan implementation and the major factors contributing to nonconformity is essential if the field of environmental planning is to be considered a legitimate policy making endeavor. Despite theoretical controversy and methodological limitations, measuring plan performance can be accomplished. In fact, monitoring the degree of plan implementation should be a routine part of every planning agency's activities. While the specific expectation for implementation should be left to the community developing the plan, some degree of accountability is warranted if plans are to become meaningful agents of ecosystem protection and more broadly sustainability.

Chapter 11

Does Planning Work? Testing the Implementation of Local Environmental Planning in Florida

This chapter tests the concepts of conformity debated in the previous chapter through two related case studies in Florida. These cases measure, map, and model the degree to which development patterns over time adhere to the original spatial design of local land use plans. Spatial and statistical analyses seek to address the issue of whether planning works as a growth management tool in Florida, which contains one of the strongest local planning mandates in the US. The first case uses GIS and associated spatial analytical techniques to compare original adopted local plans with subsequent development as indicated by wetland alteration permits. The second case relies on spatial statistical modeling to identify the major factors driving nonconforming development patterns in the southern portion of the state. Both case studies test the strength of plan implementation as a means for improving the effectiveness of planning in the future. They seek to answer the fourth research question posed in Chapter 1: what motivates the implementation of plans and policies over the long term?

CASE 1: Plan Quality versus Plan Conformance: Examining the Spatial Pattern of Wetland Development Permits Over a Ten-year Period

The first case tests the effectiveness of comprehensive planning and plan implementation across Florida by examining the spatial pattern of wetland development permits over a ten-year period. As part of the statewide comprehensive planning mandate, local jurisdictions in Florida must identify areas designated for growth to guide future development, reduce negative environmental, social, and economic impacts, and provide adequate public services to community residents. Comprehensive plans and associated future land use maps are thus the regulatory and prescriptive growth management policy instruments used by local jurisdictions. Despite the importance of local plan adoption as a legally binding growth management tool, the success of their implementation has never been systematically examined.

We address this issue by spatially comparing the original land use design of comprehensive plans in Florida with subsequent development activity. Specifically, we identify spatial clusters of wetland development permit activity and evaluate these locations against the adopted future land use maps for all county and city jurisdictions across the state. Through this analytical approach, we can find answers to the following questions: 1) How and where have wetlands been developed over a ten year period (1993-2003); 2) Are wetland permits clustered in areas designated for high density development (conformity) or do they deviate significantly from the plan's original spatial design (nonconformity); and 3) Does the quality and content of the original plan based on environmental and plan implementation policies relate to its degree of implementation based on our measurement of plan conformity? Our conclusions provide insights into how to test the effectiveness of plans as development guidance tools and how to improve the degree to which plans are implemented at the local level in Florida and in other states with planning mandates.

The Sample

As already mentioned, Florida requires that each local community prepare a legally binding comprehensive plan. Under this state mandate, comprehensive plans must adhere to the goals of the State Plan, follow a consistent format (in terms of production, element types, and review/updating processes), and most importantly provide a blueprint for future city and county growth patterns. Rule 9J-5, adopted by the Department of Community Affairs (DCA) in 1986, requires that specific elements and goals be included in local plans and prescribes methods local governments must use in preparing and submitting plans. At the heart of this coercive and highly detailed state-planning mandate lies the requirement for each local jurisdiction to adopt a future land use (FLU) map. This "regulatory and prescriptive" map designates the types of land uses permitted in specific areas within each local jurisdiction. The requirement is meant to ensure that growth and development proceeds with adequate public infrastructure, does not adversely impact critical natural habitats (e.g. wetlands), and does not promote the harmful effects of urban and suburban sprawl.

Each adopted plan under the state mandate is thus a regulatory policy instrument offering spatial guidance for future development patterns. It is not just a broad, strategic policy statement, but a set of explicit directives adopted through a participatory planning process where future outcomes are expected conform to the original design of the plan. While this so-called "blueprint" approach to planning has been heavily criticized in the past, it offers an ideal opportunity to test the degree to which development outcomes adhere to the adopted plan and indicate precisely where significant deviations may occur.

To test the degree of plan implementation, we selected all available state and federal permits issued (under part IV of chapter 373 of Florida Statutes and section 404 of the Clean Water Act) to alter a wetland in Florida between 1993 and 2002 by ecosystems defined as watershed. We used watersheds to select and summarize permit data because it is a functional ecological unit within which wetlands are located. When examining the effectiveness of plan implementation based on wetland alteration, it is appropriate to focus on areas within ecological boundaries as opposed to those defined by humans, such as local jurisdictions (Williams et al., 1997). We therefore examined approximately 39,960 issued wetlands permits within 52 adjacent watersheds as defined by the US Geological Service's (USGS) fourth order Hydrological Unit Code (HUC) (Figure 11.1). This hydrological unit is considered the most appropriate scale for assessing and implementing watershed approaches to management. We also selected a sub-sample of 1,640 wetland clusters (described below) in the southern portion of the State to examine the relationship between policies within local comprehensive plans and the degree of plan conformity.

Measuring Nonconformity

To determine the degree to which wetland development permits conform to the original design of comprehensive plans, we selected a statewide-digitized coverage of future land use for all city and county jurisdictions in Florida. This dataset was created in 1992 by the Southwest Florida Regional Planning Council which compiled each of the state's 11 regional planning councils' future land use maps, gathered from 458 local governments. Because land use categories can vary by local jurisdiction, they were placed into one of following ten classes to derive a standardized map for the entire state: Agriculture, Single Family, Estate, Multi-Family, Commercial/ Office, Industrial, Mining, Military, Preserve, and Water bodies. This future land use coverage provided a basis for evaluating the degree of conformity of wetland development permits between 1993 and 2003.

Figure 11.1 Florida watersheds
Source: Brody et al., 2005: 163.

The degree of plan conformity was measured based on several spatial analytical steps conducted in a GIS framework. First, we used the original township range (for which the data was originally organized) to total the number of permits over the study period. The State of Florida is divided into 54,285 township-range units, with an average size of 2.6 square kilometers. This procedure enabled us to calculate an intensity variable with which to conduct spatial statistical analyses across multiple watersheds. Second, we used a measure of spatial autocorrelation to identify and map significant hotspots or clusters of permits granted across the study area. These clusters represent adjacent townships containing a large number of permits (high values surrounded by high values) and indicate where intense levels of development occurred in each watershed. To locate these hotspots of high-density wetland development, we calculated a local indicator of spatial autocorrelation (LISA) (Anselin, 1995). This procedure allowed us to identify and map the statistically significant clusters of issued permits. LISA's detect significant spatial clustering around individual locations and pinpoint areas that contribute most to an overall pattern of spatial dependence. We used a local Moran's I statistic given by:

$$I_i = \frac{(Z_i - \overline{Z})}{S_z 2} * \sum_{j=1}^{N} \left[W_{ij} * (Z_j - \overline{Z} \right] \qquad (1)$$

where $\overline{Z}$ is the mean intensity over all observations, Z_i is the intensity of observation i, Z_j is intensity for all other observations, j (where $j \neq i$), S_Z^2 is the variance over all observations, and W_{ij} is a distance weight for the interaction between observations i and j.

Third, we reclassified the future land use data layer into two values: conforming and nonconforming. Conformity occurs when high-density development occurs in areas previously designated for such events. We conservatively measured conforming areas as clustered permits granted in areas designated for growth. These include Single Family, Multi-Family, Commercial/Office, Industrial, Mining and Military land uses. Nonconformity takes place when dense development is located in areas not intended by the spatial design of the originally adopted plan. Nonconforming areas were measured by combining land use designations meant for low density or no development. These include Agriculture, Estate, and Preserve land use designations.

Fourth, the spatial clustered permits data layer was overlaid on top of the reclassified data layer of future land use to determine the degree to which clusters were conforming or nonconforming. The percentage of area for each cluster containing nonconforming values was calculated to derive a measure for conformity on a scale of 0-1, where 0 is completely conforming and 1 is completely nonconforming. For example, as shown in Figure 11.2, if the land use pattern in a clustered townships was 10 percent Commercial, 30 percent Single Family, and 60 percent Preserve, then the township would receive a score of .60 nonconforming. While we expected comprehensive plans and their future land use maps to be updated and modified over the study period, spatial changes are almost always minor and a complete reversal of land use intent (e.g. from preserve to industrial) is even more of a rarity.

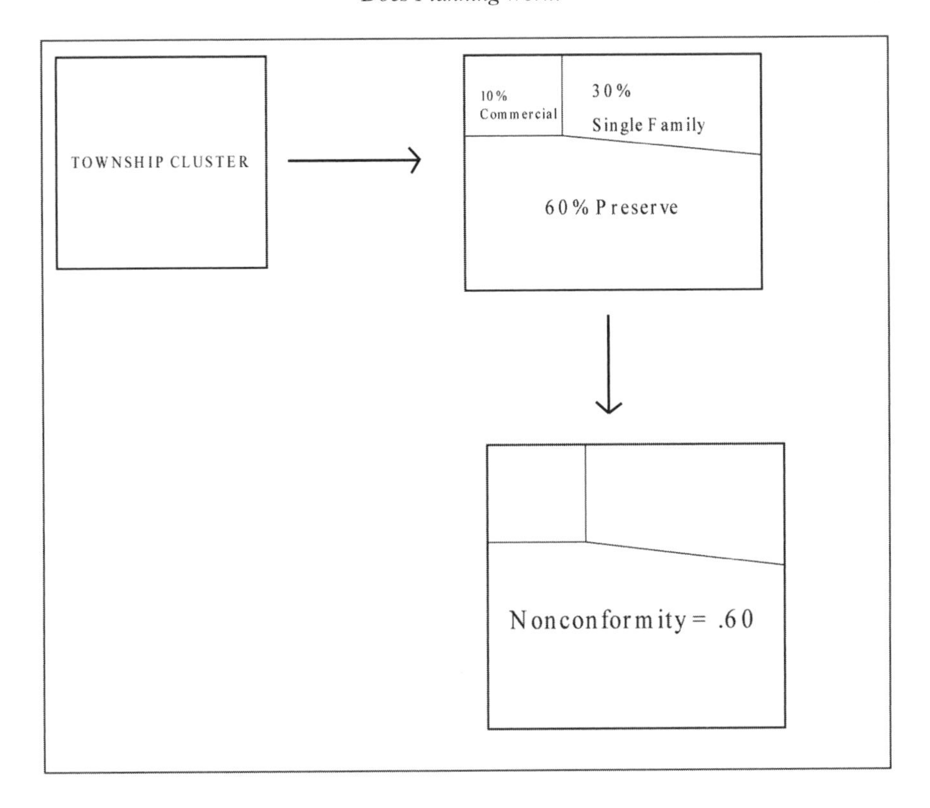

Figure 11.2 Measuring nonconforming townships
Source: Environmental Planning and Sustainability Research Unit, Texas A&M University.

Measuring Plan Quality

Plan quality was measured by evaluating the comprehensive plan for each jurisdiction occupied by a significant wetland permit cluster. Policies within the plans (plan quality indicators) were categorized into the following two major components: environmental policies and plan implementation. Environmental policies are general guides to decisions (or actions) about the location and type of development to ensure that plan goals are achieved (Berke and French, 1994). We selected and evaluated from the plan coding protocol (development in Chapter 4) each local plan for the presence of seven policies that are considered effective planning tools for concentrating growth while protecting critical habitats such as wetlands (Duerksen et al., 1997; Beatley, 2000). These policies are likely to help guide growth in an ecologically sustainable manner and assist local communities in attaining the intended spatial design and land use intensities designated in their plans. The absence of such policies may allow for more sprawling development patterns involving an increasing loss of wetlands and leading to a greater degree of nonconformity. Selected environmental policies include: use restrictions in and around critical habitats, density restrictions in and around critical habitats, targeted growth areas away from sensitive habitats/critical areas, capital improvements programming to protect critical habitat and ecological processes, density bonuses in exchange for habitat protection, transfer development rights (TDRs) away from critical habitats and clustering away from habitat and/or wildlife corridors.

The plan implementation component represents a commitment to implementing the final plan in the future (but does not indicate how well the plan is actually implemented once it is adopted). An important attribute of a high quality plan is that it articulates mechanisms and procedures to implement the plan once it is adopted. Implementation depends not only on the ability of a community to implement its plan in a timely fashion, but also on designating responsibility for actions, enforcing adopted standards, and applying sanctions to those who fail to comply. This plan component also focuses on monitoring both ecological conditions and plan effectiveness. Specific plan quality indicators thus include: clear designation of responsibility for implementation (accountability), sanctions for failure to implement regulations specified, clear timetable for implementation, regular update procedures and plan assessments, enforcement of habitat or ecosystem protection, provisions for technical assistance, monitoring for ecological processes critical habitat and indicator species, monitoring of ecological and human impacts, identification of costs or funding for implementation, monitoring of plan effectiveness, and monitoring of policy response to new scientific information. Through these 11 indicators a community can most effectively adapt to changing conditions by setting updated standards to obtain stated goals and objectives.

An environmental policy or plan implementation mechanism was coded if it was intended to protect ecologically significant habitat and restrain sprawling development that would adversely impact additional wetlands (see Table 11.1 for a complete listing of indicators selected from the original coding protocol). As done in previous case studies in this book, each indicator was measured on a 0-2 ordinal scale, where 0 is not identified or mentioned, 1 is suggested or identified but not detailed, and 2 is fully detailed or mandatory in the plan. In addition to recording the presence of each plan indicator, we calculated a plan quality index for each plan component, placing the plan component on a 0-10 scale.

Emergence of Wetland Development Clusters

Figure 11.3 illustrates an increased number of issued permits across Florida between 1993 and 2002. Generally, the number of granted wetland development permits increased significantly during the study period. 2,487 permits were granted in 1993 while approximately 4,796 permits were granted in 2002. A significant spike in the number of permits granted occurred between 1994 and 1995, representing a possible increase in the level of statewide development activity during that year. The number of permits actually declined each year from 1998 to 2000, and then sharply increased from 2000 to 2003. Since these are statewide totals, it is difficult to determine exactly what drove the yearly variations in the number of permits, but these fluctuations should be further investigated in subsequent studies.

Table 11.1 Selected plan quality indicators

Plan quality indicators
Environmental policies
Use restrictions in and around critical habitats
Density restrictions in and around critical habitats
Targeted growth areas away from sensitive habitats/critical areas
Capital improvements programming to protect critical habitat and ecological processes
Density bonuses in exchange for habitat protection
Transfer development rights from critical habitats
Clustering away from habitat and/or wildlife corridors
Implementation policies
Clear designation of responsibility for implementation
Sanctions for failure to implement regulations specified
Clear timetable for implementation
Regular update procedures and plan assessments
Enforcement of habitat or ecosystem protection
Provisions for technical assistance
Monitoring for ecological processes critical habitat and indicator species
Monitoring of ecological and human impacts
Identification of costs or funding for implementation
Monitoring of plan effectiveness and
Monitoring of policy response to new scientific information

Source: Adapted from Brody et al., 2005: 165.

Interestingly, the area of spatially clustered permits (as recorded by townships) for each year follows a similar pattern as with the number of permits (Figure 11.4). The area of clustered permits increased from 2,387 square kilometers in 1993 to 4,069 square-kilometers in 2002. Yearly fluctuations in clustered area generally match those for the total number of issued permits. This result may indicate that wetland development occurred in a relatively dense spatial configuration or in concentrated areas as opposed to randomly scattered across the state.

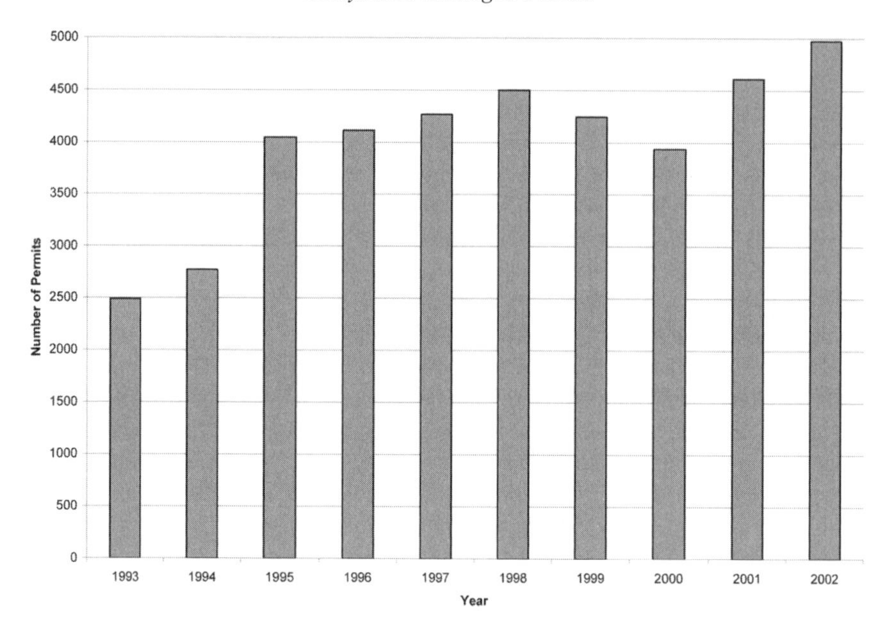

Figure 11.3 Number of wetland permits granted between 1993 and 2002 in Florida

Source: Brody et al., 2005: 166.

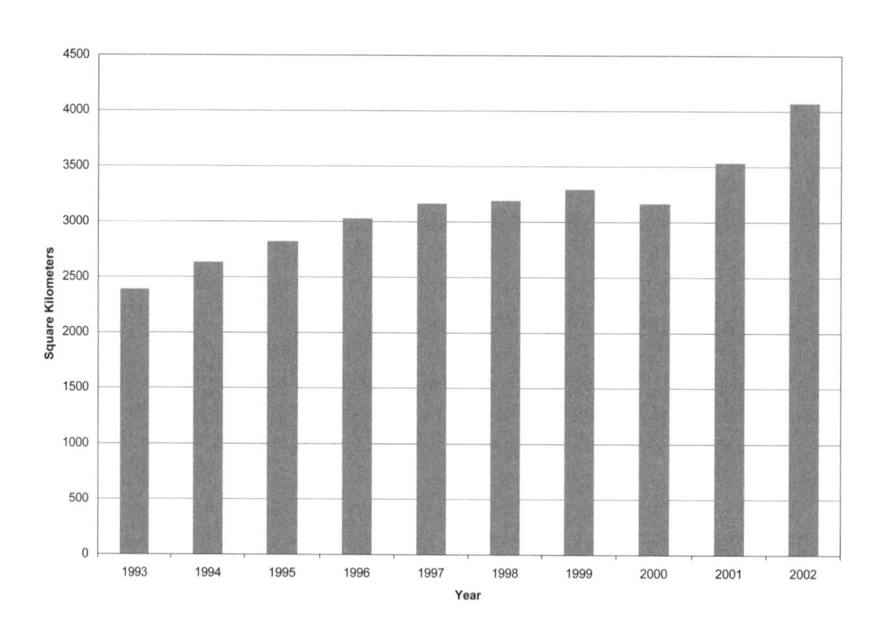

Figure 11.4 Area of clustered wetland permits between 1993 and 2002 in Florida

Source: Brody et al., 2005: 167.

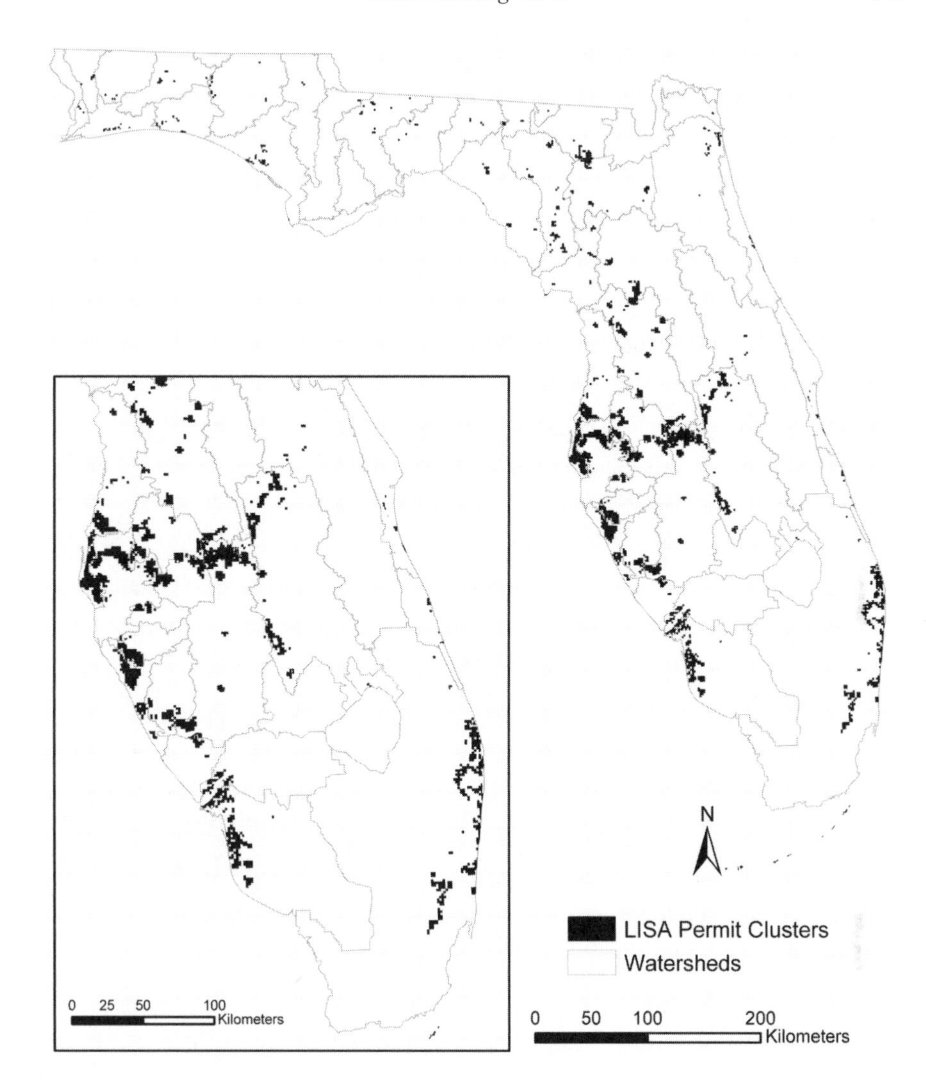

Figure 11.5 Spatial clusters of wetland permits
Source: Brody et al., 2005: 168.

Figure 11.5 illustrates the significant clusters or hotspots (showing only areas which have a LISA category of high values surrounded by high values) of issued permits that emerged at the end of the study period. Several important observations can be made based on the spatial pattern of permit clusters. Principally, the majority of clusters are located in the southern portion of the state and along the coastlines. Hotspots are particularly evident in the southeast urban corridor from Miami north to West Palm Beach (Lower East Coast Watershed), in the Southwest Coast Watershed from Naples north to Bradenton, and in and around Pinellas County/Tampa Bay Watershed. A large cluster of wetland development is also noted in the central part of the state west of Tampa Bay within the Kissimmee River Watershed. These hotspots of wetland development activity appear to mimic the general pattern of development that occurred in Florida in the 1990s and early 2000s. That is, residential and tourism development built upon and expanded outward from previously established urban centers in coastal areas in the southern portion of the state. Sprawling development into the interior areas was constrained by the presence of the Everglades National Park and Big Cyprus Preserve in the extreme south, but less so in areas north of Lake Okeechobee where there is no protective barrier.

Level of Conformance for Wetland Permit Clusters

The emergence of spatial clusters of wetland development permits and their location provides a backdrop for unraveling what may be a more pertinent question: do these clusters representing concentrated wetland alteration conform to the general design of the local planning framework? Table 11.2 shows the degree to which clusters in each watershed in the study area adhere to future land use designations established in 1992. We calculated an average conformance score for each watershed as well as the percentage of clustered area within each quartile on the conformance scale ranging from 0 (completely conforming) to 1 (completely nonconforming).

Overall, the 1st quartile (where the conformance score is equal to or less than .25) contains the most area, approximately 3,500 km^2, suggesting that the majority of wetland development across the state is relatively in conformance with the spatial intent of local plans. However, the 4th quartile, where development conformance is the lowest, contains approximately 800 km^2, which is more than the second and third quartiles combined. In fact, more than 15 percent of all clustered wetland development permits are more than 75 percent nonconforming based on the future land use maps of their associated comprehensive plans.

The worst performing watersheds (nonconformity is greater than 50 percent of all permit clusters) are located in the northern part of the state, particularly in the Panhandle region. Among these 11 watersheds, Nassau, Escambia, and Alapaha Watersheds are entirely nonconforming in their development of wetlands. In contrast, the best performing watersheds with the highest level of conformity are primarily located in the southern portion of the State along the coastlines. These areas contain the majority of population and urban centers such as Miami, Tampa, and Fort Lauderdale. While it appears from a watershed perspective that wetland development patterns in northern Florida disregard planning initiatives, these clusters actually represent small, isolated instances of local developments. The vast majority of issued permits and significantly clustered area occur in the southern portion of the state. In fact, the 11 worst performing watersheds amount to only 9 percent of all clustered area identified in Florida.

Table 11.2 Watershed plan conformance scores

Watershed	Area, Square Kilometers	Average Non Conforming Percent	Area <.25	Percent of Total	Area <.50	Percent of Total	Area <.75	Percent of Total	Area < 1.0	Percent of Total
Pensacola Bay	24.34	0.000	24.34	1.000	0.00	0.000	0.00	0.000	0.00	0.000
Choctawhatchee Bay	29.41	0.001	29.41	1.000	0.00	0.000	0.00	0.000	0.00	0.000
Chipola River	5.13	0.002	5.13	1.000	0.00	0.000	0.00	0.000	0.00	0.000
St Johns River, Upper	21.09	0.032	21.09	1.000	0.00	0.000	0.00	0.000	0.00	0.000
Taylor Creek	3.79	0.039	3.79	1.000	0.00	0.000	0.00	0.000	0.00	0.000
Sarasota Bay	236.99	0.047	221.47	0.935	10.68	0.045	2.24	0.009	2.60	0.011
East Coast, Middle	15.58	0.070	13.87	0.890	1.71	0.110	0.00	0.000	0.00	0.000
St Andrews Bay	62.02	0.085	51.70	0.834	5.20	0.084	5.08	0.082	0.03	0.000
Caloosahatchee River	138.70	0.093	128.65	0.928	0.00	0.000	6.04	0.044	4.02	0.029
Tampa Bay	333.45	0.094	307.52	0.922	18.16	0.054	2.68	0.008	5.09	0.015
Crystal River To St.	429.86	0.102	374.77	0.872	41.41	0.096	3.53	0.008	10.16	0.024
Indian River, South	15.28	0.103	12.75	0.835	2.53	0.165	0.00	0.000	0.00	0.000
Hillsborough River	246.88	0.112	209.66	0.849	15.46	0.063	14.92	0.060	6.83	0.028
Everglades-West Coast	339.44	0.124	287.70	0.848	12.99	0.038	17.66	0.052	21.09	0.062
Little Manatee River	56.71	0.142	43.60	0.769	10.47	0.185	0.00	0.000	2.65	0.047
Econfina-Fenholoway	43.49	0.170	31.79	0.731	8.06	0.185	2.61	0.060	1.02	0.024
St Johns River, Lower	56.69	0.180	29.03	0.512	1.25	0.022	25.39	0.448	1.03	0.018
Withlacoochee River, North	14.76	0.200	9.54	0.647	0.00	0.000	0.00	0.000	5.22	0.353
New River	1.11	0.214	0.23	0.209	0.88	0.791	0.00	0.000	0.00	0.000

Table 11.2 continued

Watershed	Area, Square Kilometers	Average Non Conforming Percent	Area <.25	Percent of Total	Area <.50	Percent of Total	Area <.75	Percent of Total	Area < 1.0	Percent of Total
Manatee River	144.58	0.223	103.32	0.715	8.21	0.057	5.36	0.037	27.69	0.192
Keys	24.61	0.235	12.84	0.522	7.10	0.289	1.30	0.053	3.37	0.137
Peace River	531.90	0.236	354.93	0.667	57.87	0.109	47.28	0.089	71.81	0.135
Santa Fe River	100.70	0.242	66.40	0.659	7.87	0.078	2.90	0.029	23.52	0.234
Southeast Florida	634.78	0.265	384.57	0.606	89.35	0.141	51.70	0.081	109.1	0.172
Kissimmee River	390.38	0.298	235.95	0.604	44.01	0.113	44.55	0.114	65.87	0.169
Alafia River	174.94	0.303	105.17	0.601	19.89	0.114	23.64	0.135	26.24	0.150
Perdido Bay	7.80	0.313	5.21	0.668	0.00	0.000	0.00	0.000	2.59	0.332
Withlacoochee River, South	306.77	0.340	170.63	0.556	36.27	0.118	34.83	0.114	65.05	0.212
Suwannee River, Upper	76.71	0.369	45.14	0.589	0.00	0.000	0.00	0.000	31.56	0.411
Myakka River	104.97	0.371	62.32	0.594	13.00	0.124	0.00	0.000	29.65	0.282
East Coast, Upper	13.78	0.386	8.22	0.597	0.00	0.000	0.00	0.000	5.56	0.403
St Marks River	10.41	0.388	5.22	0.501	0.00	0.000	2.62	0.252	2.57	0.247
Blackwater River	10.08	0.442	4.67	0.463	0.00	0.000	0.00	0.000	5.41	0.537
Charlotte Harbor	38.65	0.452	15.75	0.408	4.27	0.110	2.63	0.068	16.00	0.414
Yellow River	23.24	0.541	7.73	0.333	2.59	0.111	5.12	0.220	7.80	0.336
Waccasassa River	44.16	0.561	14.92	0.338	0.00	0.000	5.25	0.119	24.00	0.543
Apalachicola River	5.69	0.600	0.53	0.094	0.00	0.000	0.00	0.000	5.16	0.906
Suwannee River, Lower	135.39	0.607	37.01	0.273	15.61	0.115	12.19	0.090	70.59	0.521

Table 11.2 continued

Watershed	Area, Square Kilometers	Average Non Conforming Percent	Area <.25	Percent of Total	Area <.50	Percent of Total	Area <.75	Percent of Total	Area < 1.0	Percent of Total
Ochlockonee River	39.17	0.637	10.39	0.265	2.63	0.067	2.64	0.068	23.50	0.600
Oklawaha River	152.36	0.724	18.46	0.121	17.11	0.112	33.87	0.222	82.91	0.544
Aucilla River	10.54	0.747	2.70	0.256	0.00	0.000	0.00	0.000	7.85	0.744
Choctawhatchee River	29.10	0.825	0.00	0.000	8.38	0.288	0.00	0.000	20.72	0.712
Alapaha River	2.59	1.000	0.00	0.000	0.00	0.000	0.00	0.000	2.59	1.000
Escambia River	12.89	1.000	0.00	0.000	0.00	0.000	0.00	0.000	12.89	1.000
Nassau River	3.70	1.000	0.00	0.000	0.00	0.000	0.00	0.000	3.69	1.000
Totals	5105.0	15.000	3478.1	0.681	462.96	0.091	356.03	0.070	807.5	0.158

Source: Adapted from Brody et al., 2005: 169.

Small pockets of nonconforming development are important indicators of the effectiveness of local planning and should not be overlooked. However, a thorough examination of the value of land use designations, spatial guidance for future growth, and plan implementation should also focus on where the most intense development is taking place: southern Florida. A closer look at the conformance level of wetland development permit clusters based on quartiles (Figure 11.6) reveals an interesting spatial pattern. Nonconforming clusters occur at the fringes of coastal urban areas where development pressures are the greatest. The nonconforming patches are almost always located adjacent to conforming development. These areas include the western outskirts of Miami, Boca Raton, and West Palm Beach on the southeast coast and areas to the east of Bradenton and Sarasota on the west coast of the state. As mentioned above, areas to the north of Lake Okeechobee in the central part of the state do not have large protected areas to constrain growth and therefore contain significant clusters of wetland permits. Large patches of nonconformance are located around urban growth areas associated with Disney World just south of Ocala and the Kissimmee River. Based on the observed patterns of nonconforming wetland development, it appears urban areas in southern Florida (surrounding the Everglades Ecosystem) have experienced unintended growth towards interior portions of the state, causing critical wetlands to be filled in for development. As development pressure increased, urban and tourism areas tended to push outward and were, in this case, only constrained by large nationally protected areas.

It is still difficult to determine if local comprehensive planning has mattered statewide in terms of focusing development and protecting wetland habitat since we cannot compare conformance patterns in Florida both with and without a planning mandate. Indeed, the spatial configuration of development and the level of nonconformity might have been very different in the absence of regulatory and prescriptive land use plans.

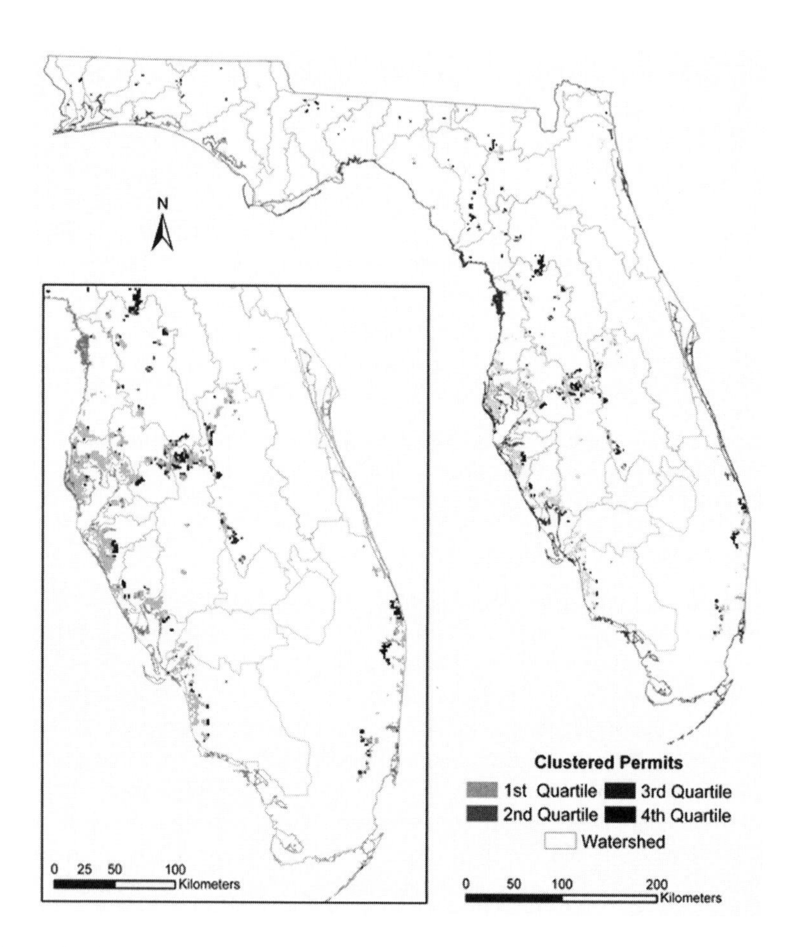

Figure 11.6 Level of conformance for spatial clusters of wetland permits
Source: Brody et al., 2005: 171.

Correlating Plan Quality with Plan Conformance

Another way we assessed the effectiveness of comprehensive planning is to examine the relationship between plan content and plan outcome. In the final phase of analysis, we conducted zero-order correlation analysis for permit clusters located in the southern part of Florida between plan quality indicators and the plan conformity measure. As shown in Table 11.3, both the environmental policy and plan implementation components are not significantly correlated with the conformity measure. That is, the presence of environmental policies and implementation mechanisms in the sample of local plans does not guarantee plan conformity when it comes to containing the development of wetlands in areas designated as undesirable. Note that Berke et al. (2006) found a different pattern of conformity in New Zealand.

Table 11.3 Correlations between plan quality indicators and plan conformity

Correlations between environmental policies and plan conformity	
	Nonconformity
Environmental policy component	0.04
Critical habitat protection	0.074**
Density restrictions	0.060*
Targeted growth	0.161**
Protection with capital improvements	-0.110**
Density bonuses	0.075**
Clustered growth	-0.092**
Transfer of development rights	0.080**
Correlations between implementation indicators and plan conformance	
	Nonconformity
Implementation component	0.067**
Designation of responsibility for implementation	0.144**
Technical assistance identified	0.006
Funding for implementation outlined	0.154**
Sanctions for failure to implement regulations	-0.066**
Timetable for implementation	0.032
Regular update procedures specified	0.176**
Enforcement of ecosystem protection specified	-0.024
Monitoring for ecological processes	0.081**
Monitoring for human resource use/impacts	-0.048*
Monitoring specified for plan effectiveness	-0.104**
Monitoring specified for policy response to new scientific information	0.101**

Source: Adapted from Brody et al., 2005: 172.

Note: n = 1640; **Correlation significant at the 0.01 level; *Correlation significant at the 0.05 level.

However, unpacking the indices and examining each individual indicator illustrates a different picture. Several policy indicators appear to correlate with significant increases in the degree of subsequent nonconformity such as the protection of critical habitat, targeted growth strategies, density bonuses, and transfer of development rights. In contrast, two policies in the index significantly increase the level of plan conformance by reducing spatial nonconformity. Wetland protection using capital improvements programming and clustered growth requirements both increase spatial conformity ($p<.01$).

Specific indicators for plan implementation reveal a similar pattern when correlated with plan conformance. Designation of responsibility for implementation, suggested funding mechanisms, requirements for regular plan updates (required under the State mandate), provisions for monitoring ecological processes, and monitoring specified so that jurisdictions can respond to new information, all significantly decrease plan conformity. On the other hand, strict sanctions for failure to implement required policies and monitoring plan effectiveness both appear to significantly increase plan conformance ($p<.01$ level). Also, monitoring human impacts on the natural environment (i.e. water quality, habitat fragmentation, storm water runoff, etc.) increases conformity ($p<.05$).

Does Planning Work to Protect Wetlands?

The results of this case study are mixed on whether local planning in Florida is protecting naturally occurring wetlands at the watershed unit. First, wetland development, as indicated by state and federal issued permits, has increased steadily between 1993 and 2002, particularly in the southern portion of Florida. The area of permit clusters followed the same upward trend, indicating that, on average, development did not occur haphazardly across the state, but in specific or concentrated areas. As communities grow and expand outward, new developments tend to locate near previous ones rather than as isolated patches outside an urban center. This cumulative spatial development pattern, so characteristic of rapidly growing communities, may help explain why, as the number of issued permits increased, the area of clustered permits also increased.

Second, the degree to which spatial clusters of wetland development permits conform to the original spatial design of local plans varies across watersheds and between the northern and southern portions of the state. While the highest levels of nonconformity are located to the north in the Panhandle region, coastal areas to the south by far contain the largest number of permits and area of nonconformity. We explain this result by the occurrence of two types of wetland development: 1) small, isolated patches in the comparatively undeveloped Panhandle and 2) rapidly expanding development in the south that pushes into the fringes of urban areas containing large populations. The presence of protected areas associated with the Everglades ecosystem in the south appears to act as a growth barrier that confines development to coastal areas. Third, the relationship between plan quality and plan conformity in southern portions of the state showed mixed results. Capital improvements programming and clustered development to protect critical habitat are the most influential policies related to plan conformity. Sanctions for failure to comply with regulations, monitoring human impacts on the integrity of the natural system, and monitoring the effectiveness of the plan itself are the implementation mechanisms most closely associated with plan conformance.

By using spatial and statistical analyses to measure the degree of plan conformance, this case study provides a stronger understanding of the link between plan making and plan implementation. The value of our approach is that it provides a spatial compass for keeping a plan on track and ensuring effective implementation over the long term. By offering a baseline with which to evaluate the effectiveness of implementation we are able to geographically isolate deviations from the original plan and potential adverse impacts to wetland systems. While the desirability of development patterns should be a value-based assumption made by a community, this method at least helps planners recognize when and where there is nonconformity and a significant change in direction from original plan design. It serves as a statistical and graphic tool with which to gauge the direction of plan implementation, adjust course to updated information, or chart a new heading before negative outcomes become irreversible. If employed by local planners, such a system could facilitate an adaptive approach to regional growth and environmental management where communities can make micro-adjustments more informally and more often than the usual official seven year plan update cycle. An adaptive approach to long-term planning can more effectively mitigate undesirable outcomes such as sprawl and environmental degradation or prevent development patterns from taking major detours from the originally intended path.

CASE 2: Planning At the Urban Fringe: Factors Influencing Nonconforming Development Patterns in the Southern Part of Florida

Case 2 builds directly on the previous case in which we found nonconforming areas of development to be spatial indicators of urban and suburban sprawl. For example, nonconforming clusters occur at the fringes of coastal urban areas containing large populations where development pressures are most intense. The next step in a thorough examination of the degree to which plans are implemented subsequent to adoption is to identify the factors driving nonconforming development patterns. This line of inquiry should help explain why development may or may not adhere to the original spatial design of the plan and provide guidance to planners both within Florida and in other states on how to mitigate nonconforming development (or sprawl) in the future. Specifically, this case study seeks to answer the question: what are the major physical, socioeconomic, and market-based factors influencing the degree of development conformity in southern Florida?

The Sample

We used the same unit of analysis, methods to derive the sample of ecosystems, and techniques for measuring nonconformity as in Case 1 of this chapter, but focused on the southern portion of Florida where development has been most intense. This sub-sample involved examining approximately 36,350 issued wetlands permits within 20 adjacent watersheds as defined by the USGS fourth order Hydrological Unit Code (HUC) (Figure 11.7).

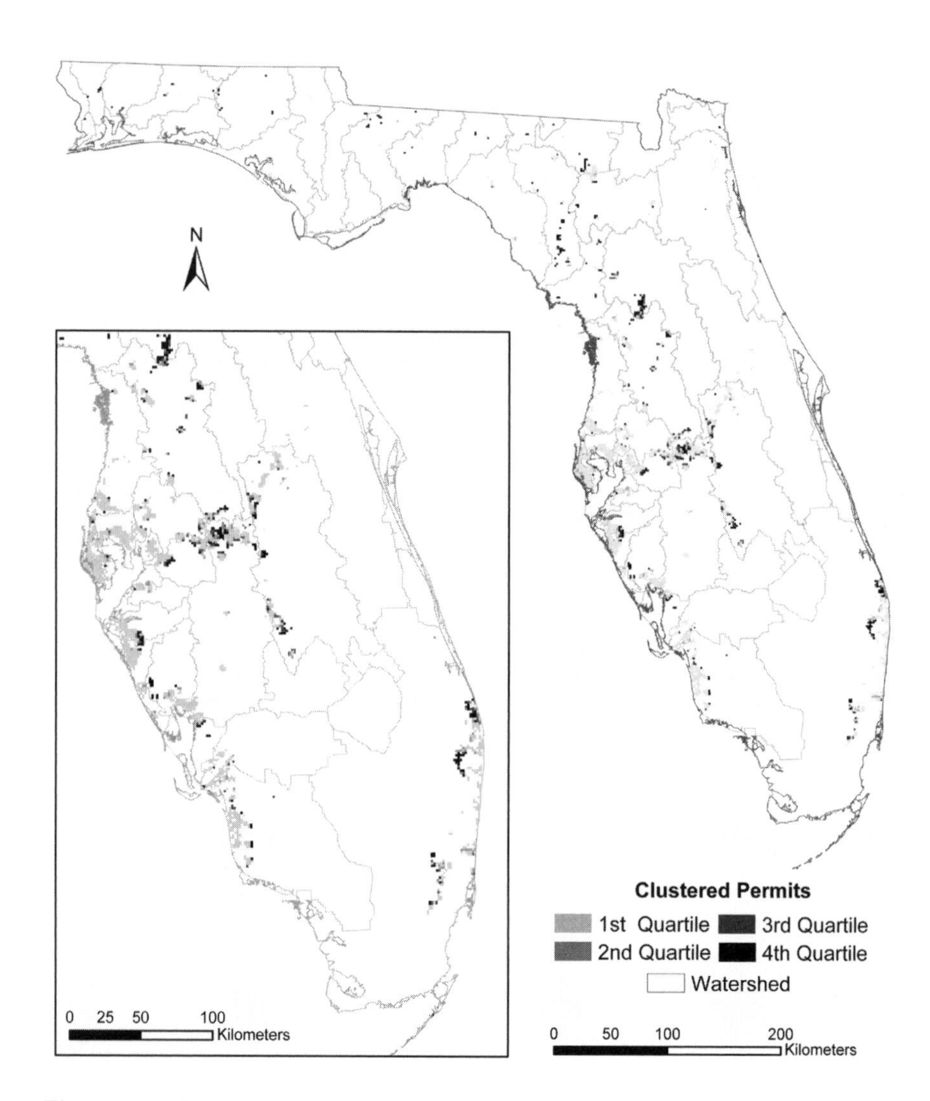

Figure 11.7 Study area
Source: Brody et al., 2006: 88.

Measuring Factors Influencing Nonconformity

To model nonconformity, we measured and analyzed the following three suites of independent variables based on the literature presented in Chapter 10: geographic variables, socio-demographic variables and market/policy variables.

Four separate geographic variables were selected to help explain nonconformity in south Florida: distance to protected areas, distance to major roads, distance to the coast, and distance to 1990 Census Places. Four socio-demographic variables were also calculated and analyzed in a statistical model. These variables included: median household income, proportion of the minority population, proportion of the population over 50, and population density. Due to the small size of the township-rang, exact populations for each unit were not feasible. Instead, we used the 1990 US Census Bureau's TIGER block group summary level to spatially transfer population estimates from each block group to the township-range unit of analysis. In cases where a township-range crossed two or more block groups, we used the average.

Land values from 1992 county tax records were used to calculate a total land value for each township-range based on specific land uses. An environmental policy index was measured by evaluating the comprehensive plan for each jurisdiction occupied by a significant wetland permit cluster. We evaluated each local plan for the presence of 4 policies (taken from the original plan coding protocol developed in Chapter 4) that are considered effective planning tools for concentrating growth while protecting critical habitats such as wetlands (Duerksen et al., 1997; Beatley, 2000; Brody et al., 2006a). Environmental policies include: capital improvements programming to protect critical habitat and ecological processes, density bonuses in exchange for habitat protection, transfer development rights (TDR's) away from critical habitats and clustering away from habitat and/or wildlife corridors.

As done in previous analyses, an environmental policy (or plan quality indicator) was coded if it was intended to protect ecologically significant habitat and restrain sprawling development that would adversely impact additional wetlands. Each indicator was measured on a 0-2 ordinal scale, where 0 is not identified or mentioned, 1 is suggested or identified but not detailed, and 2 is fully detailed or mandatory in the plan. Under the assumption that not one, but a set of policies working together in a plan facilitates conforming development, we calculated an environmental policy index based on the three steps. First, the scores for each of the indicators (I_i) were summed within each of the plan components. Second, the sum of the scores was divided by the total possible score for each plan component ($2m_j$). Third, this fractional score was multiplied by 10, placing the plan component on a 0-10 scale. That is,

$$PC_j = \frac{10}{2m_j} \sum_{i=1}^{m_j} I_i \qquad (2)$$

where PC_j is the plan quality for the jth component, and m_j is the number of indicators within the jth component.

What is Driving Sprawl?

Results from spatial regression analysis (Table 11.4) indicate a spatial lag of 10 miles has a highly significant impact on the dependent variable, plan conformity ($p<.000$).[1] That is, nonconforming development clusters are spatially dependent within 10 miles of each other and that analyzing a model that does not incorporate a spatial lag (i.e. OLS regression) may result in biased parameter estimates and misinterpretation of relationships between x and y variables. Land values are also significantly correlated with the degree of planning conformity ($p<.000$), where high values are located in areas of conforming development, primarily in urban areas. In contrast, low land values are associated with nonconforming development where residential and commercial projects have pushed into outlying rural and conservation areas.

Table 11.4 Factors influencing nonconforming development*

Variable	Coefficient	Std. error	z-value	Probability
Protected area distance	-1.26E-05	2.41E-06	-5.239	0.000
Road distance	0.000120934	1.37E-05	8.845	0.000
1990 Places distance	1.51E-05	4.38E-06	3.437	0.000
Coast distance	4.12E-07	2.98E-07	1.385	0.166
Median household income	1.29E-06	5.07E-07	2.551	0.011
Proportion minority	-0.03309794	0.03588156	-0.922	0.356
Proportion over 50	-0.07654848	0.03647497	-2.099	0.036
Population density	-0.000822687	0.000411903	-1.997	0.046
Environmental policy index	-0.00293748	0.003338019	-0.880	0.379
Land values	-1.70E-05	3.21E-06	-5.299	0.000
Lagged nonconformity	0.6771092	0.03657364	18.514	0.000
Constant	0.1362462	0.0383495	3.553	0.000
R-squared = 0.362149		Log likelihood = -109.502		
Sigma-square = 0.0659512		Akaike info criterion: 243.005		
S.E of regression = 0.257		Schwarz criterion: 307.425		
n = 1585		Degrees of Freedom: 1573		

Source: Adapted from Brody et al., 2006: 90.

Note: *A Global Moran's I statistic for the dependent variable (nonconformity) indicated significant spatial autocorrelation. This result led us to model major factors influencing nonconforming development using a spatial autoregressive model (SAR) consisting of a maximum likelihood estimation (MLE) with a spatial lag variable on the right hand side of the equation.

---

1 Because township clusters analysed in this study are not always adjacent, it was necessary to define an appropriate lag distance in order to specify a spatial weights matrix. While the determination of lag distances can often be subjective, we relied on a common practice which examines the spatial pattern of a major variable influencing the variation on the

Proximity variables are also important factors driving the degree of nonconforming development, complimenting the findings for land values. Distance from the nearest major road is the strongest predictor, where development in close proximity to highways and other primary arterials significantly increases conformity to the spatial design of local plans. In contrast, development farther away from roadways increases the likelihood that wetland development will be nonconforming. Distance from major protected areas also significantly impacts the degree of plan conformity based on the location of wetland alteration permits. Intense development activity occurring further away from protected areas such as Big Cyprus and the Everglades tends to be more conforming. This result supports our previous speculation that protected areas act as a buffer for sprawling or nonconforming growth in Florida and can help confine growth to the urban core (Brody et al., 2006b). Finally, proximity to settled populations where public services such as sewer and water are most likely available has significant implications for local plan conformity. Development close to or within a settled area is more conforming. In contrast, wetland development clusters located on the periphery of commercial and residential centers where public infrastructure is less likely is an indicator that development patterns have deviated from the original intent of the adopted plan.

Socioeconomic and demographic variables in the statistical model have less of an impact on plan conformity compared to market-based and geographic factors. Wealthy residents, as measured by median home values, are associated with significantly greater degrees of nonconformity ($p<.01$). This result reflects a common pattern of development in Florida where large homes are built in planned subdivisions (often gated) away from urban centers. These planned developments attract relatively wealthy second homeowners and seasonal tourists from out of the State. Those attracted to resort-oriented residential communities originally designated for rural land uses are most likely young in age. While the effect is fairly weak, the percentage of the population over 50 years of age are associated with greater degrees of plan conformity ($p<.05$). High population density is associated with increased plan conformity, although we would expect a more statistically significant effect considering the greatest concentration of people should be located in the urban core, rather than outlying suburban and ex-urban communities.

Finally, it is important to note from a planning perspective that environmental policies have a negative, but non-significant effect on the degree of plan conformity. In other words, even when policies meant to reduce sprawl and increase spatial conformity are adopted in local comprehensive plans, they do not appear to significantly increase the likelihood that development will adhere to the original spatial design of the plan itself.

---

dependent variable. The literature on development described in the previous chapter highlights land value as the most important factor influencing development patterns, providing us with a rationale to observe the spatial pattern of this variable to specify the spatial lag. Mapping land values in urban areas, where the majority of the wetland clusters are located, revealed a clear break in land value intensity approximately 10 miles from a city center. From this analysis, we concluded a ten mile lag distance defines "neighbors" as all nonconforming township-range units within ten miles of each other based on centroid-to-centroid Euclidian distance.

Holding the Line in the Face of Development Pressures

This case suggests that the majority of wetland development clusters in southern Florida reasonably conform to the original spatial design of local land use plans. At the same time, a significant portion of these clusters is over 75 percent nonconforming, particularly where development is accelerating into the outskirts of urban cores. This sprawling pattern of growth, where residential development occurs in areas previously designated for agricultural use or conservation necessitates a planning focus on the fringe of urban areas. To mitigate high degrees of nonconformity (>75 percent), which can lead to adverse environmental, social, and economic impacts, planners and other public decision makers must orient growth management policies and programs towards the ever-fading transition zone between urban and rural areas. This domain is where planners must hold the line in the face of development pressures that can encroach on critical natural resources and agriculture operations. A focus on the urban fringe may include, among other alternatives, local planning strategies such as greater restrictions on wetland development, a sharper distinction between urban and rural areas through the designation of Urban Growth Boundaries (UGBs), incentives that promote clustered development and higher densities in the urban core, careful placement of public facilities and capital investment, and programs that encourage infill development or redevelopment in central urban areas.

Spatial statistical modeling indicates there are several factors impacting the degree of nonconforming wetland development, each with distinct planning implications:

First, nonconforming wetland development clusters are significantly spatially correlated up to ten miles apart (tests for spatial autocorrelation were not performed for greater distances). This result suggests that, on average, a dense area of wetland development does not stand-alone in space, but occurs in relatively close proximity to other development clusters. Additionally, the formation of one cluster will encourage others to emerge in the same general area. Understanding this pattern of development visually and quantitatively is important for planners interested in mitigating sprawl and unintended outbreaks of nonconforming development. For example, permitting a large-scale development project in a previously designated rural area can become a catalyst for future development nearby, even when limited public facilities are available and local growth management policies have been adopted. Making project level decisions without regard to the broader spatial ramifications may, over time, promote unintended patterns of development.

Second, the value of land strongly contributes to the degree of plan conformity. Residential developers are often eager to purchase comparatively inexpensive property outside of urban areas originally containing wetlands or agricultural operations. Just as higher profit margins attract developers, more affordable housing prices in locations away from the congestion of cities appeal to prospective homebuyers, particularly seasonal residents. This phenomenon is driven by what Mattson (2003) calls rising "trigger levels." The trigger level is defined as the point within the development process when a combination of declining agricultural prices, rising public service costs, and increased local property tax assessments cause an urban-rural fringe property owner to sell his or her land. By selling, the landowner perpetuates the occurrence of sprawl and unintended development outside of urban areas.

Third, proximity to likely public services, potential recreational areas, and major transportation corridors significantly affects the degree of plan conformity. These geographic variables support the visual results described above: that nonconforming wetland development occurs on the fringe of urban centers and far from essential public infrastructure. This trend can be interpreted in different ways. On the one hand, development adhering to the spatial design of the local plan is close to major roadways, water treatment facilities, and away from ecologically sensitive protected areas. Since the majority of clustered area leans toward conformity (< 25 percent), there is evidence that planners are effectively placing public infrastructure in designated growth areas while preventing development from encroaching on critical natural resources. On the other hand, the most nonconforming development clusters occur primarily outside of urban centers, suggesting that even the most well-intentioned spatial planning designs can not guarantee conformity or prevent the adverse impacts of sprawling growth patterns.

Fourth, wealthy homeowners appear to be driving nonconforming development through preferences for newly constructed resort communities located outside of congested downtown areas. This trend facilitates the development of large single-family homes often situated on golf courses where wetlands once predominated. While southern Florida will continue to be an attractive resort and retirement destination, planners should encourage developers to build communities that adhere to "smart growth" or "New Urbanist" principles and are situated closer to urban centers. Such options include planning policies, such as urban growth boundaries, clustering of development, and mixed use zoning, among others. Additionally, financial incentives including special tax districts, transfer of development rights programs, and density bonuses can help persuade developers to locate their projects within existing urban or commercial areas. Projects such as Seaside in the Panhandle region and Myzner Place on the south east coast provide lifestyle alternatives that reverse the trend of nonconformity discussed above, but are relative anomalies compared to most large-scale developments across the state.

Finally, planning policies such as those mentioned above that promote a well-defined urban core and reduce sprawling growth patterns are clearly not enough by themselves to ensure conforming development. This finding is evidenced by the fact that the environmental policy index analyzed in the spatial regression model was not statistically significant. In addition to strong plans and policies, implementation mechanisms need to be adopted, such as accountability, enforcement, sanctions for failure to comply, and perhaps most importantly participation of key stakeholders in the planning process. As demonstrated by numerous studies (Burby, 2003; Brody, 2003; Brody et al., 2003 to name a few), public participation increases ownership over and accountability for the contents of a plan, often leading to stronger levels of implementation.

PART 4
Planning Implications and Recommendations

Results from the multiple case studies presented throughout this book show that existing human management and planning systems are not functioning in a sustainable fashion. This inability to effectively maintain the integrity of natural systems over the long term in part results from a spatial and temporal disconnect between ecological systems and human perceptions, decision-making, and collective action. The key to effective ecosystem planning, and ecologically sustainable development in general, is to reduce and eliminate this disconnect. First, we need to better understand how to gear management and policy to the ecological unit as opposed to one defined by human boundaries. Second, we need to better understand how to facilitate proactive approaches to management as opposed to reactionary responses to environmental crises. This is not solely a technical or engineering problem, but one that involves addressing the complex interaction of human decisions and the biophysical environment.

By examining local comprehensive plans, this book provides key insights into how to effectively accomplish ecosystem management in Florida and other areas across the US. Five years of research on this topic has made several contributions to the practice of managing ecological systems for several reasons. First, developing a conceptual and measurable model of a high quality local ecosystem management plan moves the field of environmental planning away from qualitative assessments of plan quality toward an evaluative technique that is more precise, defensible, and comparable across multiple jurisdictions. Understanding exactly what makes a strong local ecosystem management plan provides practitioners with a model against which to test the effectiveness of existing plans and policies. Second, demonstrating the extent to which local jurisdictions are managing natural systems in Florida provides insight into how to strengthen existing planning frameworks. Identifying the relative strengths and weaknesses in local management statistically and spatially across the State helps planners improve policies to more effectively protect critical natural resources over the long term. Florida can serve as a guiding example for other states and regions across the Country. Third, understanding the major influences on ecosystem plan quality not only tests important theoretical assumptions about environmental planning (e.g. stakeholder participation), but also provides information about how to construct planning processes to protect biodiversity and associated natural systems. Finally, demonstrating how plan implementation associated with ecosystem management can be spatially monitored and quantitatively assessed offers insights on how to make a good plan "stick" over the long run.

The findings of this study can thus guide practitioners in the US and in other countries in improving their capabilities to manage ecological systems in the future. Part 4 of this book sets forth a series of *proactive planning levers* designed to incorporate ecosystem considerations into plans and planning processes before substantial degradation takes place. These recommendations are based directly on the empirical findings of this book. They attempt to answer the fifth and final research question: how can plans, planning processes, and the state growth management programs that mandate them be improved to enhance ecosystem management?

Chapter 12

Recommendations for Improving the Process and Practice of Environmental Planning at the Local Level

The following recommendations based on the findings of this book may assist planners in Florida and other regions to incorporate the principles of ecosystem management into local plans and policy instruments. These recommendations aim to facilitate a proactive approach to natural resource management, rather than to institute policies long after adverse human impacts have taken place. Most importantly, they provide direction to planners on how, from a spatially bottom-up perspective, the integrity, functions, and processes of ecological systems can be protected over the long term. Recommendations are categorized as the following: the plan, the process, and implementation.

The Plan

Improve the Factual Basis of the Plans: The first step in increasing the overall quality of a local plan is to improve its factual basis by conducting a more thorough resource inventory and incorporating available data on existing natural resource conditions. A strong factual basis helps a community understand what resources are being adversely impacted or are in need of further protection. With a greater understanding of existing critical resources, planners and planning participants may be more likely to incorporate ecosystem management policies at the outset of adverse human impacts.

As previously mentioned, the factual basis ranks as the lowest scoring plan component in our analyses and leaves the most room for improvement. Incorporation of ecosystem components, such as identification of keystone species, areas of high biodiversity, and habitat corridors will help a community better implement the principles of ecosystem management. For example, jurisdictions can make use of the Florida Fish and Wildlife Conservation Commission's digital maps of focal species, areas of high biodiversity, and habitat conservation areas. These maps could be analyzed in combination with existing land use patterns to help identify potential conservation zones.

Increase the Use of Geographic Information Systems: Geographic Information Systems (GIS) technology is a powerful tool to both display and analyze natural resource data. It helps planners not only understand precisely where critical habitats exist but the degree to which they are in need of protection. As an analytical tool, GIS helps project the future and enables planners to make proactive choices about the management of existing natural resources. GIS also can serve an educational function by explaining complex problems to planning participants who are not technically oriented.

There are hundreds of GIS data layers available to local jurisdictions throughout Florida ranging from watershed boundaries to vegetation cover. However, only a few communities in the study sample take advantage of the large amounts of free existing data and the analytical power of this technology in making ecologically sustainable planning choices. For example, only 7 percent of the 30 jurisdiction random sample in the study incorporated Gap Analysis data layers in their plans. Planning offices do not need to hire technical personnel or purchase expensive equipment to successfully use GIS in planning. Data layers easily can be downloaded in several formats from state or regional organizations.

Increase Monitoring Activities: It is important not only to identify existing natural resources, but also to understand how baseline conditions change over time. Monitoring ecological processes, critical habitats, and the impacts to these resources from human activities plays an essential role in anticipating the decline of ecosystems and setting preventative policies. Managers must be able to react to constantly changing ecological systems, sudden shifts in interests and objectives, and a continuous barrage of new and often ambiguous information. A strong local monitoring program can provide a powerful informational lever for identifying adverse impacts to biodiversity before they become irreversible.

The majority of the jurisdictions studied in this book designated monitoring programs, primarily related to water quality. However, it is unclear how data from monitoring will be fed back into the decision-making process and enable the plan to act as a flexible policy instrument. Through monitoring, jurisdictions can most effectively practice adaptive management, a continuous process of action-based planning, monitoring, researching, and adjusting with the objective of improving future management actions (Holling, 1995; Endter-Wada et al., 1998). For example, jurisdictions can initiate a community based water-monitoring program for coastal estuaries. Changes in nutrient levels can be reported to the local planning or environmental agency, which then can take action can be taken before major declines in water quality threaten to fisheries or recreational areas.

Generate More Specific Goals and Policies: One of the major weaknesses of the plans examined in this book is their lack clear directives and specific ecosystem management goals and policies. Descriptions of programs or specific actions often are overly vague and diffuse. Plans need more specifics, particularly for goals, to guide the implementation of ecosystem management initiatives. Clear and detailed goals often have timelines when they must be accomplished. Strong objectives can be measured or have measurable targets (i.e. a 40 percent reduction in nitrogen run off). For example, the goal "to protect natural systems" comes across as vague and difficult to interpret. On the contrary, the goal "manage and enhance viable native ecological communities to protect the functions of natural systems and the diversity of native plants, animals, and fisheries, particularly those endangered or threatened" is much more specific and effective at generating strong policies. Similarly, an objective to "reduce nonpoint source pollution or nutrient run off into estuaries" is far less effective than an objective to "reduce the levels of nitrogen and phosphorous entering Tampa Bay by 25 percent by the year 2005."

Expand the Planner's Toolbox: The plans examined in this study concentrate primarily on a narrow set of regulatory actions, such as land use restrictions or conservation zoning. However, the use of incentive-based policies in plans, such as density bonuses, transfer of development rights, and preferential tax treatments can be more effective in achieving the goals of ecosystem management at the local level. Most importantly, such policies encourage rather than force parties to protect critical habitats and areas of high biodiversity. For example, allowing increased densities for residential developments in exchange for the protection of critical wetland habitat enables developers to meet their objectives while instilling motivation to protect important ecological components. Efforts to protect ecosystems become more proactive when landholders act because they want to, not because they have to. In this way, incentive-based strategies encourage community members to think about and act on the principles of ecosystem management before they must be coerced with a regulatory "stick."

Initiate Environmental Educational Programs: Education is the most profound way to change behavior and generate proactive ecosystem management practices. Local outreach programs can build public awareness on the importance of protecting the value of critical natural resources and maintaining ecological integrity. Educational strategies include informational workshops, information dissemination (printed and electronic), presentations, and community programs such as monitoring or waste cleanup. Learning through involvement fosters a sense of place and facilitates action to protect the natural environment upon which communities depend. For example, citizen beach cleanup programs have been extremely effective in fostering environmental awareness in Fort Lauderdale and other coastal jurisdiction across the state. Only half of the random sample evaluated in this book includes public environmental education programs in its set of policies, indicating that the link between planning and education is being underemphasized. Yet, those jurisdictions that incorporated environmental education policies generated significantly higher quality plans ($p<.01$).

Perhaps the most important group to target environmental education programs is private sector landowners. It is imperative that the benefits of collaborative ecosystem management are clearly articulated to this stakeholder since it has the largest impact on biodiversity in Florida and throughout the US. If industry and major land developers can comprehend the long-term economic benefits of collaboration when it comes to the sustainable management of their own resources, the potential media attention, gains from sharing data and information, and finally the future gains of personal relationships and networks, the goals of ecosystem management in Florida will be more easily attained.

More Effectively Transfer Ecosystem Management Programs to the Local Level: At the state level, Florida possesses one of the most ambitious ecosystem management programs in the country. Since 1993, the DEP has been committed to implementing the principles of ecosystem management or the newly termed "watershed management" across the state. However, local jurisdictions have not effectively incorporated these concepts or initiatives into their plans and programs. None of the plans sampled even mention the presence of an EMA plan or project within their jurisdiction. There remains a disconnect between the state and local government levels that is hampering the ability of communities to manage ecological systems. Ecosystem efforts at the federal level are also not well incorporated into local plans. For example, National Estuary Programs (NEP) are rarely discussed or supported within local plan elements. To implement the principles of ecosystem management at the local level, there must be a more efficient transfer of ideas from state and federal levels of government to local jurisdictions. For example, local jurisdictions could easily include the policies of an EMA or NEP plan to ensure that regional efforts take place at the local level. Sarasota was able to fulfill many of its environmental goals by adopting the Tampa Bay NEP plan in their comprehensive plan. The DEP and other state organizations could facilitate local level commitment by providing a greater degree of technical assistance or educational outreach to ensure that their programs filter down to the local level where they may have the greatest impact.

Spatially Target Program Implementation: From a strategic perspective, if state and regional planners know that by boosting the quality of one jurisdiction's plan, it is likely that the plans of surrounding jurisdictions will also be enhanced, they can focus limited time, personnel, and financial resources on a single jurisdiction as a way to improve the management of a larger ecologically defined area. Thus, rather than implementing ecosystem programs across broad regions, state managers may consider pinpointing one or a few jurisdictions as a strategy to achieve regional management. Additional study is needed to determine if the strengthening of one plan leads to a domino effect in plan quality for surrounding jurisdictions. Analysis of the social, political, and economic relationship between specific jurisdictions is necessary to more thoroughly understand adjacency issues. This study simply identifies that an interface between jurisdictions may exist, not what specifically happens at this interface.

The Process

Target Key Stakeholders for Participation in the Planning Process: A key recommendation stemming from the results of this book is that planners should target key stakeholders for participation during the planning process. One of the most statistically powerful findings is that the presence of certain stakeholders, particularly industry, significantly increases local ecosystem plan quality. As previously described, when organizations bring to the planning process valuable knowledge of critical habitats and innovative ideas of how to sustainably manage these habitats based on their own experience, it can strengthen the ability of the final plan to achieve the principles of ecosystem management. Planners must recognize the specific contributions each stakeholder can make and aggressively target these groups for participation throughout the planning process. A strategy of targeted participation can make certain the stakeholders that have the most to contribute are present during the planning process. Targeted participation can, however, become a balancing-act because some groups will favor one issue but not another.

Less than 20 percent of the 30 jurisdiction sample targeted (as opposed to actually included) any type of resource-based industry group for participation in the planning process. In contrast, 60 percent of the sample targeted local business groups, such as storeowners, and approximately half targeted neighborhood associations. In this sense, industrial stakeholders represent an untapped planning resource that has the ability to boost the collective capacity of planning participants, resulting in a stronger, better balanced plan that not only meets the interests of the community, but is more likely to be implemented over time.

Industries such as marine, forestry, and agricultural groups are usually left out of the planning process, despite the fact that much of the critical habitat in the US is located on private lands. Industry often times conducts its own monitoring activities and maintains large databases on environmental conditions. By sharing information throughout the planning process, this stakeholder group has the ability to boost the collective capacity of those drafting the plan, leading to more effective management of ecological systems at the local level. The greatest opportunity to achieve collaborative management of entire ecological systems that transcend organizational boundaries thus lies with the participation of industry in the decision-making process.

If a comprehensive plan has the potential to directly or indirectly affect industry lands, industry collaboration throughout the planning process is essential. Recognizing the importance of industry participation, both Pinellas and Sarasota Counties directly targeted and involved agricultural and marine interests throughout the planning process. These organizations were able to contribute valuable knowledge of existing natural resources that resulted in more focused policies in the adopted plan. Of the different types of industry groups analysed in this book, forestry and marine groups seem to have the strongest impact on raising the quality of the final plan. Since these groups rely on a healthy and ecologically productive natural environment for economic success, they have a vested interest in maintaining the overall integrity of the natural systems. In contrast, agricultural groups, particularly in the southern portions of the state, have traditionally opposed all types of environmental management initiatives. Planners may want to target specific types of industry that will be the most likely to enhance the planning process and the final plan.

Based on the findings of cases in previous chapters, environmental NGOs also contribute to significantly raising the quality of a local ecosystem plan. In addition to bringing to the planning process essential data on environmental conditions and expertise on environmental management techniques, these groups help raise interest and commitment to protecting critical components of ecosystems over the long term. Environmental NGOs, such as the Audobon Society and the Nature Conservancy, often act as both technical consultants and advocacy-based groups. They can play a critical role in the planning process by raising awareness of environmental issues related to planning and focusing participants' attention on protecting the function of natural systems. Environmental NGOs thus boost technical planning capacity, as well as ensure that environmental matters are included in community discussions.

Take Proactive Steps to Reduce Negative Impacts of Rapid Development: Local comprehensive planning is intended to serve as a proactive policy-making process where communities lay out their vision of development patterns and conservation initiatives well into the future. A central issue for local watershed planning thus becomes how to motivate communities to protect critical ecosystem components before they are severely impacted by human growth and development. Cases in this book show that even with a strong state local planning mandate, communities in Florida are not proactively planning for ecological integrity. Careful monitoring of regional development trends and potential associated negative impacts to critical natural resources is a starting point for stimulating the adoption of plans to protect ecosystem components early in the process of natural resource decline. Regional monitoring of both the human and natural environment can serve as an early warning system which invokes a proactive approach to management. Once potential "train wrecks" are identified, state-level organizations can put the financial and personnel related resources into place to accommodate watershed planning in the face of rapid growth and development.

Fortifying local planning capacity in concert with environmental education programs can facilitate ecologically sustainable approaches to development before major environmental impacts occur. Although proactive approaches to local planning may require the commitment of time and resources at the outset, the long-term investment should be profitable considering the exorbitant costs of ecological restoration, removal of invasive species, and improvement of water quality. As evidenced by the studies presented in this book, waiting for the necessary planning capacity and public interest to materialize along with human disturbance associated with rapid growth and development may not be the most effective strategy to manage ecological systems in Florida. Matching planning agency capacity with the level of expected regional growth could trigger ecosystem-planning initiatives before adverse environmental impacts take place.

Development in the Panhandle of northwest Florida provides an example of the ecological hazards of reactionary approaches to planning. St. Joe Paper Company owns approximately 1,000,000 acres of the coastal Panhandle, which until recently was a relatively undeveloped region containing some of the last unprotected areas of high biological diversity in the State. In 1998, the Company launched its Northwest Florida strategy, which now includes 20 separate developments and permits for over 10,000 high-priced homes (Pittman, 2003). Local planning agencies are unprepared for St. Joe's effort to recreate northwest Florida. These organizations lack the necessary staff, expertise, and ecological data to thoroughly understand the ecological implications of such expansive residential growth. Furthermore, small, once rural communities are using outdated plans and inadequate factual bases to accommodate large-scale regional development projects. As evidenced by the results of case studies in this book, waiting for the necessary planning capacity and public interest to materialize along with human disturbance associated with rapid growth and development may not be the most effective strategy to sustainably manage ecological systems in northwest Florida. Matching planning agency capacity with the level of expected regional growth could trigger ecosystem-planning initiatives before adverse environmental impacts take place.

More Clearly Articulate the Benefits of Industry Participation: Planners and public decision makers need to better engage resource-based industries in local environmental planning processes by articulating the benefits of participation for the corporations themselves. First, participation in collaborative ecosystem management offers an attractive alternative to command-and-control style government regulation and could reduce the need for strict regulatory controls in the long run. For example, in a national study on forestry companies and ecosystem approaches to management, most companies expressed their dislike of stringent regulations and lack of control in the regulatory process (Brody et al., 2006). The practice of collaborative ecosystem management could help decrease the burden of governmental controls and improve the current regulatory structure by educating regulators about sustainable corporate practices. Furthermore, a collaborative approach may reduce operational costs and provide the flexibility corporations need to manage their natural resource base.

Second, participation in collaborative ecosystem management and other sustainability projects can result in a positive public image and indirect financial gain over the long term. Companies are increasingly more receptive to media coverage and the expectations of their stakeholders. As indicated in the forestry company survey, pressure exerted through these outside channels may be a viable policy option for influencing corporate decisions. Engaging in high profile, environmentally sustainable practices often results in favorable media attention and broad public support. Positive press can reduce public opposition to commercial harvesting operations, increase the firm's customer base, and make it easier to conduct core business practices.

Third, involvement in collaborative ecosystem management projects may provide a strategic opportunity to develop partnerships with other stakeholders. As previously mentioned, firms increasingly recognize that they lie within a broader network of interests and that interaction with these outside interests is essential to effective management (Hoffman, 2000). Developing relationships based on trust and reciprocity with neighboring landholders can help a company attain its resource management and financial goals. By forming relationships with other interests, there is a good chance that those interests will collaborate with each other to reach common goals in the future. Strong partnerships can also reduce the likelihood that costly and protracted disputes will emerge among multiple interests within the ecological region. Reciprocity is particularly important for corporate landholders whose neighbors are controlling and impacting what can often be considered the same natural system.

Finally, the formation of partnerships creates the possibility of information sharing, data collection, and technical assistance. Collaborating with outside parties often entails an exchange of information and data relevant to managing natural resources. Corporate entities can gain valuable knowledge regarding habitat locations, species movement, the presence of pollutants, etc. Furthermore, nongovernment organizations and government agencies can provide technical assistance and databases that can be useful to managing critical natural resources on industry owned lands.

Implementation

Identify the Spatial Pattern of Physical Development: It is important to consider not only the amount of wetland development taking place, but specifically where this development is occurring across natural landscapes. The function and integrity of watersheds depend on a patchwork system of interconnected wetlands. Some of these patches may be more important in supporting the ecological system than others. Analyzing how the spatial pattern (e.g. location, proximity, clustering) of development affects critical ecological components is thus an important aspect of environmental planning. The recent ubiquity of GIS and spatial analytical techniques provide the technical means to help local and regional planners better understand the impacts of development trends.

Focus Planning Efforts on the Urban Fringe: Because an outward expansion of growth away from urban areas in south Florida comprises the majority of nonconforming wetland development, a planning focus on the urban fringe is necessary to limit sprawling development patterns that adversely impact the Everglades ecosystem. A focus on the urban fringe may include, among others, local planning strategies such as greater restrictions on wetland development, a sharper distinction between urban and rural areas through the designation of Urban Growth Boundaries (UGBs), incentives that promote clustered development and higher densities in the urban core, careful placement of public facilities, and programs that encourage infill development or redevelopment in central urban areas.

Establish Protected Areas as a Buffer to Growth: Designation of protected areas in key locations may control rapidly expanding growth or focus development in ecologically desirable areas. While wetland development in south Florida expanded into the fringes of urban areas towards the interior of the State, it was restrained by the presence of Everglades National Park and Big Cyprus National Preserve. A lack of protected areas north of Lake Okeechobee may have contributed to the spread of concentrated wetland development into central Florida. Thus, protected areas designated by state and local authorities may provide a dual role: protection of critical natural habitats that support the integrity of ecological systems and a land use planning tool that constrains and focuses growth in areas that will reduce adverse environmental impacts.

Florida already has several programs in place to acquire ecologically sensitive lands such as the Preservation 2000 Initiative and the Florida Forever program, which use a documentary stamp tax to generate $300 million annually for acquisition of conservation lands (Beatley, 2000). At the local level, Pinellas County adopted the Penny for Pinellas program consisting of a one-cent local option sales tax which piggybacks the state sales tax and applies to all sales, use, services, rentals, admissions and other authorized transactions. Proceeds from the local option sales tax can be used only for capital projects. Of this money, over $1 million was dedicated for preserves and habitat management and approximately $3.3 million for parks and land acquisition in 2003 (http://www.pinellascounty.org/Penny/default.htm). Penny for Pinellas may contribute to the fact that the Tampa Bay watershed (encompassing Pinellas County) has an average conformity score of less than 1 percent and over 90 percent of its wetland development clusters are in the first quartile of nonconformity (see Chapter 11 for more details).

Adopt Capital Improvements Policies to Increase Plan Conformity: Policies that entail capital improvements programming may be one of the most effective planning tools to ensure plan conformity and the protection of critical habitat. The presence of public infrastructure and facilities is a major catalyst for land development. Local governments can contain or guide new development by not budgeting for water or sewer lines, roads, or other types of infrastructure in certain areas (Duerksen et al., 1997). Only 15 percent of the plans analyzed in the book incorporated capital improvements programming and control of public investments as a way to protect critical natural habitats such as wetlands. More widespread use of this planning strategy may increase the degree of plan conformity and plan implementation in general.

Adopt Clustering of Development Policies to Increase Plan Conformity: Clustering development is another planning technique that is strongly associated with plan conformity. On a regional scale, clustered development patterns help contain growth within the urban core and protect critical habitats. At the parcel level, cluster zoning allows high-density development in one area of a parcel while leaving the remaining land undeveloped. This concept is widely used to contain local growth and set aside sensitive areas such as wetlands and wildlife habitat (Beatley, 1997). Clustered development may be strongly related to plan conformity in part due to its direct and easily recognizable benefits: protecting significant areas of natural habitat without negatively impacting land values.

Enforce Sanctions for Noncompliance to Increase Plan Conformity: Sanctions designated for failure to implement goals, objectives, and policies may motivate communities to conform to the original plan design and lead to a greater degree of plan performance. Although mandatory sanctions in the form of penalties, added restrictions, and requirements appeared rarely in the plans for southern Florida (10 percent), this implementation mechanism seems to trigger increased plan conformity over time. This result indicates that if there are legal or financial consequences embedded in a plan for not adhering to its requirements, communities are more likely to take planning directives more seriously.

Regularly Evaluate Plan Performance: Specific monitoring activities designated in a plan may lead to greater plan conformity and better overall plan implementation. Assessing the effectiveness of the plan itself is the most important monitoring mechanism because it forces planners and communities to continually re-assess plan performance and make adjustments based on new information or changing conditions. Regular plan updates, self-assessments, and report cards for plans are vital for keeping a plan on track. With a system of constant self-reflection on the effectiveness of an adopted plan, planners can become adaptive managers responsive to the shifting political, socioeconomic, and physical landscape. Most importantly, monitoring a plan can catch systematic occurrences of nonconformance and associated implementation failures before they become too severe. Another useful monitoring approach involves tracking human impacts on the natural environment. A clear understanding of the adverse impacts caused by urban development and resource degradation can assist planners in mitigating loss of ecosystem structure and function. When incorporated into a planning process and final plan, this information communicates the importance of protecting wetland function and integrity at the watershed level. It should be noted that some degree of monitoring and evaluation of plans does take place at the local level. As part of the state planning mandate, all jurisdictions are required to draft an Evaluation and Appraisal Report (EAR) every seven years. This document evaluates the progress made in obtaining the goals of a local government's comprehensive clan and determining if changes are needed.

Consider the Effects of Land Taxes: Given that inexpensive land appears to be one of the strongest predictors of nonconforming development, planners and other public officials must be conscious of the way they assess and tax real property. Currently, land is taxed based on its highest and best use, which tends to elevate trigger levels. Preferential tax treatments, on the other hand, can assess property based on actual current uses rather than its potential. In areas where pressure to develop in outlying areas not intended by the original plan create higher property values and tax burdens, current use assessments can provide tax relief to landholders who chose to continue to pursue agricultural, forestry, or conservation land uses (Duerksen et al., 1997). Another financial incentive approach to maintaining development conformity is the use of tax credits. In this instance, federal tax deductions are offered to a landowner who donates a portion of his or her property to a land trust as open space or an open space easement. This provision simultaneously rewards the landowner for reducing the potential development of his or her land while providing a potential buffer for sprawling development outward from the urban core.

Chapter 13

Conclusion

This book has argued that strong local level environmental and natural resource decision-making is essential for the long term management of ecological systems. While ecosystem approaches to management tend to focus on broad spatial scales, the principles and practices underlying these approaches must be implemented through local plans and planning processes. Because local policy instruments guide the scale and pattern of physical development, actions within town, city, and county jurisdictions can often protect critical natural resources more effectively than top-down state or federal protection mandates. In other words, local planning is where the "rubber hits the road" in terms of ecologically sustainable development practices.

The case studies presented and conclusions made throughout this book rest on what constitutes a high quality local plan that implements the principles of ecosystem management. We evaluate this model against multiple samples of local jurisdictions throughout Florida to identify strengths and weaknesses of these local plans, aspects of the community processes that led to their adoption, and the degree to which they are implemented over time. The research has relied on a wide variety of data and methods including remote sensing images, US Census data, GIS analytical techniques, phone interviews, and mail surveys. All of this empirical analysis is meant to provide the reader with a detailed picture of how well local jurisdictions are, alone and collectively, managing ecological systems over the long term.

In general, we find that the existing human management and planning systems are not taking care of themselves. There is a spatial and temporal disconnect between ecological systems and human perceptions, decision-making, and collective action. The key to ecosystem approaches to management is to reduce and eliminate this disconnect. Thus, as a society we need to understand how to gear management and policy to the ecological unit as opposed to one defined by human boundaries. We need to understand how to facilitate proactive approaches to management as opposed to reactionary responses to environmental crises. This is not simply a technical or engineering problem, but one that involves addressing the complex interaction of human decisions and the biophysical environment.

To summarize, key findings indicate that:

- the mosaic of local management across large watersheds has important gaps that need to be filled by policy makers to more effectively protect these ecosystems;
- the factual basis, the foundation upon which high quality plan rests, is the weakest component of the overall ecosystem plan quality measure;
- the amount of biodiversity within a local jurisdiction does not stimulate planners and community members to adopt better plans, but rather human

disturbance (i.e. buildings and pavement) is the strongest catalyst for what should be proactive protective measures;

- when certain groups, such as industry are involved in the local planning process, the final product is stronger in terms of ecosystem management capabilities; and
- development tends to conform to the original spatial design of an adopted environmental plan only under specific circumstances.

The value of the findings is that they are based on a measurable model of what makes for a high quality plan that can be systematically compared across multiple jurisdictions. Understanding statistically and spatially the degree to which local jurisdictions are managing ecological systems provides guidance for both local and regional planners on how to set future planning processes and land use policies that help reduce the decline of biological diversity and loss of ecosystem function which has cost Florida so dearly. This approach acts as a spatial compass for understanding the current state of planning across large watersheds and keeping plans in line with their original intent over the long term. By offering a baseline with which to evaluate the effectiveness of plans and their implementation, this book can help planners throughout Florida and elsewhere recognize when and where there is a break-down in planning effectiveness or significant change in direction from the original plan design.

The analytical techniques and results presented throughout the book offer insights to planning academics and practitioners on how to integrate proactive planning levers into local policy frameworks and maybe most importantly, how to take an adaptive approach to management where communities can adjust policies based on updated information or chart a new course before negative effects become irreversible. Overall, an adaptive approach to long-term planning can more effectively mitigate undesirable outcomes such as sprawl and environmental degradation and keep communities more in line ecologically sustainable approaches to development.

Appendices

APPENDIX A: ECOSYSTEM PLAN CODING PROTOCOL

Coder: Date Coded:

Title of Plan

Jurisdiction:

Organization that prepared document:

Year that plan is up to date:

Year that plan is adopted:

Fact base Coding categories: 0 = not mentioned 1 = identified but not detailed 2 = detailed			
A. Resource inventory	Code for detail	Page no. reference	Comments
1.1 Ecosystem boundaries/edges: • Mapped • Described	 M ___ D ___		
1.2 Ecological zones or habitat types: • Mapped • Classified	 M ___ C ___		
1.3 Description of ecological functions for habitat type or ecological zones	D ___		
1.4 Species Ranges: • Mapped • Described	 M ___ D ___		

1.5 Habitat corridors between wildlands that allow for natural movements of wide-ranging animals and migration of species: • Mapped • Described	 M ___ D ___		
1.6 Predicted distributions of terrestrial vertebrate species (animals) mapped	 M ___		
1.7 Areas with high biodiversity or species richness: • Mapped • Described	 M ___ D ___		
1.8 Classification of vegetation: • Species level • Community level	 S ___ C ___		
1.9 Classification of wildlife: • Species level • Community level	 S ___ C ___		
1.10 Natural vegetation cover mapped to the level of dominant or co-dominant species	 M ___		
1.11 Threatened and endangered species: • Mapped • Catalogued	 M ___ C ___		
1.12 Invasive or exotic species: • Mapped • Catalogued	 M ___ C ___		
1.13 Indicator species: • Mapped • Catalogued	 M ___ C ___		
1.14 Soils types/associations: • Mapped • Classified	 M ___ C ___		
1.15 Wetlands and riparian habitat: • Mapped • Described/inventoried	 M ___ D ___		
1.16 Climate described	D ___		
1.17 Water resources (other than wetlands): • Mapped • Described	 M ___ D ___		
1.18 Surface hydrology described:	D ___		

1.19 Subsurface (i.e. groundwater system, water circulation, subsurface geology) • Described: • Recharge areas mapped • Water budget measured	 D ___ M ___ M ___		
1.20 Marine resources: • Described: • Fisheries inventoried	 D ___ I ___		
1.21 Graphic representation of/acknowledgement of resources extending beyond town boundaries	 M ___ D ___		
1.22 Other Prominent landscape features/special resources: • Mapped • Described	 M ___ D ___		
B. Human ownership			
1.23 Ownership patterns for conservation lands mapped: • Public • Private	 M ___ M ___		
1.24 Management status identified for conservation lands (degree to which an area is managed to maintain biodiversity)	I ___		
1.25 Network of conservation lands mapped	M ___		
1.26 Distributions of native vertebrate species, groups of species or vegetation communities within network of conservation lands compared	D ___		
C. Human impacts/Problem identification Coding categories: 0 = Mentioned but no detail 1 = Vague explain 2 = Clear explain			
1.27 Human population growth	D ___		
1.28 Road density	D ___		
1.29 Fragmentation of habitat	D ___		
1.30 Rate of wetlands development	D ___		
1.31 Nutrient loading	D ___		

1.32 Water pollution	D ___		
1.33 Loss of fisheries/marine habitat	D ___		
1.34 Alteration of waterways	D ___		
1.35 Other factors contributing to a loss of biodiversity (specify)	D ___		
1.36 Value of biodiversity identified and discussed	D ___		
1.37 Major stakeholders and their interests identified	D ___		
1.38 Existing federal and state environmental regulations described (beyond town boundaries)	D ___		
1.39 Carry capacity identified/measured	D ___		
1.40 Incorporation of Gap Analysis into conservation aspects of plan	D ___		
Goals and objectives Coding categories: 0 = Not Present 1 = Present but not detailed 1 = Present			
3.1 Seek to protect the integrity of ecological systems			
3.2 Seek to protect natural and evolutionary processes and functions at the landscape level			
3.3 Seek to protect biological diversity at the landscape level			
3.4 Maintain large intact patches of native species (i.e. vegetation and wildlife) by preventing fragmentation of those patches by development			
3.5 Establish priorities for native species protection and protect habitats that constrain the distribution and abundance of those species			
3.6 Protect rare or unique landscape elements			
3.7 Protect rare and endangered species			

3.8 Maintain connections among wildlife habitats			
3.9 Represent, within protected areas, all native ecosystem types across their natural range of variation			
3.10 Seek to maintain intergenerational sustainability of natural systems			
3.11 Balance human use with the need to maintain viable populations of native species over the long-term			
3.12 Goals to restore ecosystems or critical habitat			
3.13 Other goals that further the protection and management of ecological systems (specify)			
3.14 Goals are clearly specified in that they: 1) Focus on specific regionally significant habitats/populations 2) Identify specific species 3) Include criteria for success	1) 2) 3)		
3.15 Presence of measurable objectives to achieve ecosystem management goals			
Inter-organizational coordination and capabilities for ecosystem management Coding categories: 0 = not mentioned 1 = mentioned but no detail 2 = mentioned in detail			
4.1 Other important organizations and/or stakeholders identified			
4.2 Coordination with other organizations specified to protect ecosystems or transboundary resources specified			
4.3 Coordination within the jurisdiction specified			
4.3 Intergovernmental bodies to protect ecosystems or transboundary resources specified			

4.4 Joint database production specified			
4.5 Information sharing with other organizations specified			
4.6 Link between science and policy organizations specified			
4.7 Position of the local jurisdiction within the broader bioregion identified			
4.8 Intergovernmental agreements (IGAs) specified			
4.9 Integration with other plans/ policies in the region outlined			
4.10 Conflict management processes outlined			
4.11 Commitment of financial resources specified			
4.12 Other forms of inter-organization coordination specified (i.e. communication)			
Ecosystem management policies, tools and strategies Coding categories: 0 = not mentioned 1 = suggested in plan 2 = mandatory in plan			
A. Regulatory Tools			
5.1 Use restrictions in and around critical habitats			
5.2 Density restrictions in and around critical habitats			
5.3 Restrictions on native vegetation removal			
5.4 Controls/removal of exotic/invasive species			
5.5 Buffer requirements along water courses or around critical habitats			
5.6 Controls on fencing to permit natural movement of native species			
5.7 Controls on public or vehicular access to wildlife habitat			
5.8 Phasing of development to reduce wildlife disturbance			

5.9 Controls on construction activities to protect native habitats, their associated species, and ecological processes			
5.10 Conservation zones or overlay districts to protect sensitive lands and/or wildlife corridors			
5.11 Performance zoning to reduce impacts on critical areas/habitat			
5.12 Subdivision standards to protect critical areas/ habitat/undeveloped land			
5.13 Creation of protected areas/sanctuaries			
5.14 Urban growth boundaries that do not include critical habitat			
5.15 Targeted growth areas away from sensitive habitats/critical areas			
5.16 Capital Improvements Programming to protect critical habitat and ecological processes			
5.17 Site plan review to protect habitat			
5.18 Habitat restoration actions specified			
5.19 Actions to protect resources crossing into other jurisdictions (see 4.2)			
5.20 Other regulatory tools to protect regionally significant habitats and ecosystems (specify)			
B. Incentive tools			
5.21 Density bonuses in exchange for habitat protection			
5.22 Clustering away from habitat and/or wildlife corridors			
5.23 Transfer development rights away from critical habitats			
5.24 Preferential tax treatments to protect critical habitats			
5.25 Mitigation banking			
5.26 Other incentive-based tool (specify)			

C. Land acquisition programs			
5.27 Fee simple purchase			
5.28 Conservation Easements			
5.29 Other land acquisition techniques (specify)			
D. Other policies, tools, strategies			
5.30 Control of Public Investments and Projects			
5.31 Designation of special taxing districts to raise funds for land acquisition			
5.32 Public education programs on the importance of protecting habitat and ecological systems			
5.33 Monitoring specified for: 1) Ecological processes critical habitat and indicator species 2) Human resource use/impacts			
Implementation Coding categories 0= not mentioned 1= suggested in plan 2= mandatory in plan			
6.1 Clear designation of responsibility for implementation specified (accountability)			
6.2 Provision of technical assistance identified			
6.3 Identification of costs or funding for implementation outlined			
6.4 Provisions of sanctions for failure to implement regulations specified			
6.5 Clear timetable for implementation outlined			
6.6 Regular updates and plan assessment specified			
6.7 Enforcement of habitat or ecosystem protection specified			

6.8 Monitoring Specified for: 1) Plan effectiveness 2) Policy response to new scientific information	 1) 2)		
Data presentation Coding categories 0= not present 1= Present			
2.1 Presence of digital coverages and or satellite images for: • Ecosystems • Habitat • Species	 E ___ H ___ S ___		
2.2 Use of GIS analysis	D ___		
2.3 Sources are given for background information and data	D ___		

Comments:

APPENDIX B: CONCEPT MEASUREMENT

Name	Type	Measurement	Scale	Source
Total plan quality	Dependent	Sum of five plan components: factual basis + goals and objectives + inter-organizational coordination + policies + implementation	0-50	Local Comprehensive Plans
Plan component quality	Dependent	Sum of indicator scores divided by the total possible score for that component	0-10	Local Comprehensive Plans
Item breadth	Dependent	No. of plans that address item/No. plans in sample	0-1	Local Comprehensive Plans
Item quality	Dependent	Total score of all plans that addressed an item divided by No. plans that addressed the item	0-1	Local Comprehensive Plans
Total item quality	Dependent	Item breadth + item quality	0-2	Local Comprehensive Plans
Spatially weighted plan quality	Dependent	Each jurisdiction's plan quality score is weighted by the proportion of the ecological unit occupied by that jurisdiction. Weighted scores are summed across all jurisdictions within each ecological unit	Ratio	Local Comprehensive Plans

Plan conformity	Dependent	The percentage of area for each clustered township containing nonconforming land uses	0-1, where 0 is completely conforming and 1 is completely nonconforming	Southwest Florida Regional Planning Council; Florida Geographic Data Library
Biodiversity	Independent	Area of regional biodiversity in square meters divided by the area of the jurisdiction	Interval	FWCC
Human disturbance	Independent	Area of disturbance in square meters divided by the area of the jurisdiction	Ratio	FWCC
Disturbed-biodiversity	Independent	Interaction of biodiversity and disturbance	Ratio	GIS calculation from FWCC data layer
Planning capacity	Independent	Number of planners devoted to drafting the plan	Ratio	Survey
Planning commitment	Independent	Effort devoted to protecting critical natural areas + emphasis on protecting critical habitat	Ordinal; 0-2, where 0 is no commitment and 2 is high commitment	Survey
Population	Independent	Natural log of the population estimate for a jurisdiction	Ratio	US Census
Population change	Independent	The percentage change in population within a jurisdiction between 1990 and 2000.	Ratio	US Census
Wealth	Independent	Natural log of the median home value	Ratio	US Census

Population density	Independent	The population divided by the area of the unit of analysis	Ratio	US Census
Education	Independent	Percentage of the population with a high school degree	Ratio	US Census
Income	Independent	Average Median Household Income	Ratio	US Census
Proportion minority	Independent	Average proportion of minority population	Ratio	US Census
Proportion over 50 years	Independent	Average proportion of persons over 50 years of age	Ratio	US Census
Land use	Independent	The area or proportion of area occupied by a specific land use	Ratio	FL DEP
Future land use	Independent	The area or proportion of area occupied by a specific future land use as of 1992	Ratio	Southwest Florida Regional Planning Council
Land values	Independent	Total land value of each township-range	Ratio	Florida Geographic Data Library
Stakeholder representation	Independent	Proportion of 13 possible groups participating in the planning process	Ratio	Survey
Stakeholder participation	Independent	Presence of a stakeholder in the planning process	0-1	Survey
Distance to protected area	Independent	Distance from the centroid of each township-range to the nearest protected area	Ratio	Florida Natural Areas Inventory

Distance to major roads	Independent	Distance from the centroid of each township-range to the nearest major road	Ratio	Subset of roads from FGDL functional road classification coverage
Distance to US census places	Independent	Distance from the centroid of each township-range to the nearest MSA	Ratio	US Census Bureau, 1990 TIGER coverage
Distance to coast	Independent	Distance from the centroid of each township-range to the coastline	Ratio	DEP Florida Counties Coverage

Bibliography

Chapter 1

Beatley, T. (2000), 'Preserving Biodiversity: Challenges for Planners', *Journal of the American Planning Association* 66:1, 5-20.

Burby, R. et al. (1997), *Making Governments Plan: State Experiments in Managing Land Use* (Baltimore: Johns Hopkins University Press).

Catlin, R. (1997), *Land Use Planning, Environmental Protection, and Growth Management: The Florida Experience* (Chelsea, MI: Ann Arbor Press).

Christensen, N. et al. (1996), 'The Report of the Ecological Society of America Committee on the Scientific Basis for Ecosystem Management', *Ecological Applications* 6:3, 665-691.

Cortner, H. and Moote, M. (1999), *The Politics of Ecosystem Management* (Washington, DC: Island Press).

Cox, J. et al. (1994), *Closing the Gaps in Florida's Wildlife Habitat Conservation System* (Tallahassee: Florida Game and Fresh Water Fish Commission).

Department of Environmental Protection, Florida. (1995a), Ecosystem Management Implementation Strategy, Action Plan (Tallahassee: DEP).

Department of Environmental Protection, Florida. (1995b), Ecosystem Management Implementation Strategy, Volume II – Appendices (Tallahassee: DEP).

Dobson, P. et al. (1997), 'Geographic Distribution of Endangered Species in the United States', *Science* 275, 550-553.

Duerksen, C. et al. (1997), *Habitat Protection Planning: Where the Wild Things Are*, APA Planning Advisory Report No. 470/471.

Endter-Wada, J. et al. (1998), 'A Framework for Understanding Social Science Contributions to Ecosystem Management', *Ecological Applications* 8:3, 891-904.

Grumbine, E. (1994), 'What is Ecosystem Management?' *Conservation Biology* 8:1, 27-38.

Haeubner, R. (1998), 'Ecosystem Management and Environmental Policy in the United States: Open Window or Closed Door?' *Landscape and Urban Planning* 40, 221-233.

Hoffman, A.V. (1999), 'Housing Heats Up: Home Building Patterns in Metropolitan Areas', *The Brookings Institution Center on Urban and Metropolitan Policy Survey Series* (Washington, DC: The Brookings Institution).

Kirklin, J. (1995), 'Protecting Species and Ecosystems within Planning Processes', *Environmental Planning* 12:4, 6-13.

May, P. et al. (1996), *Environmental Management and Governance: Intergovernmental Approaches to Hazards and Governance* (London: Routledge).

McGinnis, M. et al. (1999), 'Bioregional Conflict Resolution: Rebuilding Community in Watershed Planning and Organizing', *Environmental Management* 24:1, 1-12.

Noss, R., and Cooperrider, A. (1994), *Saving Nature's Legacy: Protecting and Restoring Biodiversity* (Washington, DC: Island Press).

Noss, R., and Scott, M. (1997), 'Ecosystem Protection and Restoration: The Core of Ecosystem Management', in M. Boyce and A. Hanley (eds), *Ecosystem Management: Applications for Sustainable Forest and Wildlife Resources* (New Haven: Yale University Press), pp. 239-264.

Peck, S. (1998), *Planning for Biodiversity: Issues and Examples* (Washington, DC: Island Press).

Szaro, R. et al. (1998), 'The Emergence of Ecosystem Management as a Tool for Meeting People's Needs and Sustaining Ecosystems', *Landscape and Urban Planning* 40, 1-7.

Yaffee, S. et al. (1996), *Ecosystem Management in the United States: An Assessment of Current Experience* (Washington, DC: Island Press).

Yaffee, S. and Wondolleck, J. (1997), 'Building Bridges across Agency Boundaries', in K. Kohm and J. Franklin (eds), *Creating A Forestry for the 21st Century* (Washington, DC: Island Press).

Chapter 2

Allen, T. and Hoekstra, T. (1992), *Toward a Unified Ecology* (New York: Columbia University Press).

American Forest and Paper Association (1993), *Sustainable Forestry Principles and Implementation Guidelines* (Washington, DC: American Forest and Paper Association).

Bacow, L. and Wheeler, M. (1984), *Environmental Dispute Resolution* (New York: Plenum Press).

Bingham, G. (1986), *Resolving Environmental Disputes: A Decade of Experience* (Washington, DC: Conservation Foundation).

Blumenthal, D. and Jannink, J.L. (2000), 'A Classification of Collaborative Management Methods', *Conservation Ecology* 4:2, 13.

Brunner, R. and Clark. T. (1997), 'A Practice-Based Approach to Ecosystem Management', *Conservation Biology* 11:1, 48-58.

Brussard, P. et al. (1998), 'Ecosystem Management: What is it Really?' *Landscape and Urban Planning* 40, 9-20.

Bryson, J. and Crosby, B. (1992), *Leadership for the Common Good: Tackling Public Problems in a Shared-Power World* (San Francisco: Jossey-Bass).

Carpenter, S. and Kennedy, J. (1988), *Managing Public Disputes: A Practical Guide to Handling Conflict and Reaching Agreement* (San Francisco: Jossey-Bass).

Chisholm, D. (1989), *Coordination without Hierarchy: Informal Structures in Multi-organizational Systems* (Berkley, CA: University of California Press).

Christensen, N. et al. (1996), 'The Report of the Ecological Society of America Committee on the Scientific Basis for Ecosystem Management', *Ecological Applications* 6:3, 665-691.

Coglianese, C. (1999), 'The Limits of Consensus', *Environment* 41:3, 28-33.

Conley, A. and Moote, M. (2003), 'Evaluating Collaborative Natural Resource Management', *Society and Natural Resources* 16, 371-386.

Cortner, H. and Moote, M. (1994), 'Trends and Issues in Land and Water Resources Management: Setting the Agenda for Change', *Environmental Management* 18:2, 167-173.

Cortner, H. and Moote, M. (1999), *The Politics of Ecosystem Management* (Washington, DC: Island Press).

Cortner, H. et al. (1998), 'Institutions Matter: The Need to Address the Institutional Challenges of Ecosystem Management', *Landscape and Urban Planning* 40, 159-166.

Crowfoot, J. and Wondolleck, J. (1990), *Environmental Disputes: Community Involvement in Conflict Resolution* (Washington, DC: Island Press).

Czech, B. and Krausman, P. (1997), 'Implications of an Ecosystem Management Literature Review', *Wildlife Society Bulletin* 25:3, 667-675.

Daniels, S. et al. (1996), 'Decision-Making and Ecosystem-Based Management: Applying the Vroom-Yetton Model to Public Participation Strategy', *Environmental Impact Assessment Review* 16, 13-30.

Daniels, S. and Walker, G. (1996), 'Collaborative Learning: Improving Public Deliberation in Ecosystem-Based Management', *Environmental Impact Assessment Review* 16, 71-102.

Duane, T. (1997), 'Community Participation in Ecosystem Management', *Ecology Law Quarterly* 24:4, 771-797.

Emery, F. and Trist, E. (1965), 'The Causal Texture of Organizational Environments', *Human Relations* 18, 21-35.

Endter-Wada, J. et al. (1998), 'A Framework for Understanding Social Science Contributions to Ecosystem Management', *Ecological Applications* 8:3, 891-904.

Estay, D. and Chertow, M. (1997), 'Thinking Ecologically: An Introduction', in M. Chertow and D. Estay (eds), *Thinking Ecologically: The Next Generation of Environmental Policy* (New Haven: Yale University Press).

Fisher, R. and Ury, W. (1991), *Getting to Yes* (New York: Penguin Books).

Freeman, E. (1984), *Strategic Management: A Stakeholder Approach* (Boston, MA: Pitman).

Forester, J. (1993), *Critical Theory, Public Policy, and Planning Practice* (Albany, NY: State University of New York Press).

Gerlach, L. and Bengston, D. (1994), 'If Ecosystem Management is the Solution, What's the Problem?' *Journal of Forestry* 92, 18-21.

Godschalk, D. (1992), 'Negotiating Intergovernmental Development Policy Conflicts: Practice-based Guidelines', *Journal of American Planning Association* 58:3, 368-378.

Godschalk, D. et al. (1994), *Pulling Together: A Planning and Development Consensus Building Manual* (Washington, DC: Urban Land Institute).

Gray, B. (1989), *Collaborating: Finding Common Ground for Multiparty Problems* (San Francisco: Jossey-Bass).

Gray, B. and Wood, D. (1991), 'Collaborative Alliances: Moving From Theory to Practice', *Journal of Applied Behavioral Sciences* 27:1, 3-22.

Grumbine, E. (1994), 'What is Ecosystem Management?' *Conservation Biology* 8:1, 27-38.

Habermas, J. (1979), *Communication and the Evolution of Society* (Boston, MA: Beacon Press).

Haeubner, R. (1998), 'Ecosystem Management and Environmental Policy in the United States: Open Window or Closed Door?' *Landscape and Urban Planning* 40, 221-233.

Holling, C. (1996), 'Surprise for Science, Resilience for Ecosystems, and Incentives for People', *Ecological Applications* 6:3, 733-735.

Innes, J. (1996), 'Planning Through Consensus Building: A New View of the Comprehensive Planning Ideal', *Journal of American Planning Association* 62, 460.

Jennings, P. and Zandbergen, P. (1995), 'Ecologically Sustainable Organizations: An Institutional Approach', *Academy of Management Review* 20:4, 1015-1052.

Kennedy, D. et al. (2000), *The New Watershed Source Book* (Boulder, CO: Natural Resources Law Center, University of Colorado School of Law).

Khator, R. (1999), 'Networking to Achieve Alternative Regulation: Case Studies from Florida's National Estuary Programs', *Policy Studies Review* 16:1, 65-86.

Kirklin, J. (1995), 'Protecting Species and Ecosystems within Planning Processes', *Environmental Planning* 12:4, 6-13.

Lackey, R. (1998), 'Seven Pillars of Ecosystem Management', *Landscape and Urban Planning* 40, 21-30.

Lee, R. (1992), 'Ecologically Effective Social Organization as a Requirement for Sustaining Watershed Ecosystems', in R.J. Naiman (ed.), *Watershed Management: Balancing Sustainability and Environmental Change* (New York: Springer-Verlag).

Leopold, A. (1949), *A Sand County Almanac* (New York: Oxford University Press).

Lessard, G. (1998), 'An Adaptive Approach to Planning and Decision-Making', *Landscape and Urban Planning* 40, 81-87.

Light, S. (1995), 'The Everglades: Evolution of Management in a Turbulent Ecosystem', in L. Gunderson et al. (eds), *Barriers and Bridges to the Renewal of Ecosystems and Institutions* (New York: Columbia University Press), pp. 3-34.

McGinnis, M. et al. (1999), 'Bioregional Conflict Resolution: Rebuilding Community in Watershed Planning and Organizing', *Environmental Management* 24:1, 1-12.

Norton, B. (1998), 'Evaluation and Ecosystem Management: New Directions Needed?' *Landscape and Urban Planning* 40, 185-194.

Ostrom, E. (1990), *Governing the Commons: The Evolution of Institutions for Collective Action* (New York: Cambridge University Press).

Patterson, R. (1999), 'Negotiated Developments: Best Practice Lessons from Two Model Processes', *Journal of Architecture and Planning Research* 16:2, 133-148.

Peck, S. (1998), *Planning for Biodiversity: Issues and Examples* (Washington, DC: Island Press).

Putnam, R. (1993), *Making Democracy Work: Civic Traditions in Modern Italy* (Princeton, NJ: Princeton University Press).

Randolph, J. and Bauer, M. (1999), 'Improving Environmental Decision-making Through Collaborative Methods', *Policy Studies Review* 16, 169-191.

Reicham, O. and Pullman, H. (1996), 'Scientific Basis for Ecosystem Management', *Ecological Applications* 6:3, 694-696.

Salwasser, H. (1994), 'Ecosystem Management: Can it Sustain Diversity and Productivity?' *Journal of Forestry* 92:8, 6-10.

Selin, S. and Carr, D. (2000), 'Modeling Stakeholder Perception of Collaborative Initiative Effectiveness', *Society and Natural Resources* 13, 735-745.

Selin, S. and Chavez, D. (1995), 'Developing a Collaborative Model for Environmental Planning and Management', *Environmental Management* 19:2, 189-195.

Senge, P. (1990), *The Fifth Discipline: The Art and Practice of the Learning Organization* (New York: Doubleday Currency).

Sexton, W. (1998), 'Ecosystem Management: Expanding the Resources Management "Tool Kit"', *Landscape and Urban Planning* 40, 103-112.

Shelford, V. (1933), 'A Nature Reserve Plan Unanimously Adopted by the Society', *Ecology* 14, 240-245.

Stanley, T. (1995), 'Ecosystem Management and the Arrogance of Humanism', *Conservation Biology* 9:2, 255-262.

Susskind, L. et al. (eds) (1999), *Consensus Building Handbook: A Comprehensive Guide to Reaching Agreement* (California: Sage Publications).

Susskind, L. and Cruikshank, J. (1987), *Breaking the Impasse: Consensual Approaches to Resolving Public Disputes* (New York: Basic Books).

Szaro, R. et al. (1998), 'The Emergence of Ecosystem Management as a Tool for Meeting People's Needs and Sustaining Ecosystems', *Landscape and Urban Planning* 40, 1-7.

Vogt, J. et al. (1997), *Ecosystems* (New York: Springer-Verlag).

Westley, F. (1995), 'Governing Design: The Management of Social Systems and Ecosystems Management', in L. Gunderson et al. (eds), *Barriers and Bridges to the Renewal of Ecosystems and Institutions* (New York: Columbia University Press), pp. 391-427.

Westley, F. and Vrendenburg, H. (1997), 'Interorganizational Collaboration and the Preservation of Global Biodiversity', *Organizational Science* 8:4, 381-403.

Wheatley, M. (1992), *Leadership and the New Science* (San Francisco: Berrett-Koehler Publishers, Inc).

Williams, D. and Stewart, S. (1998), 'Sense of Place: An Elusive Concept that is Finding a Home in Ecosystem Management', *Journal of Forestry* 96:5, 18-23.

Wondolleck, J. and Yaffee, S. (2000), *Making Collaboration Work: Lessons from Innovation in Natural Resource Management* (Washington, DC: Island Press).

Yaffee, S. (1996), 'Ecosystem Management in Practice', *Ecological Applications* 6:3, 724-727.

Chapter 3

Agardy, M. (1994), 'Advances in Marine Conservation: The Role of Protected Areas', *Trends in Ecology and Evolution* 9:7, 267-270.

Beatley, T. (2000), 'Preserving Biodiversity: Challenges for Planners', *Journal of the American Planning Association* 66:1, 5-20.

Chisholm, D. (1989), *Coordination without Hierarchy: Informal Structures In Multi-organizational Systems* (Berkley, CA: University of California Press).

Cortner, H. and Moote, M. (1999), *The Politics of Ecosystem Management* (Washington, DC: Island Press).

Daniels, S. and Walker, G. (1996), 'Collaborative Learning: Improving Public Deliberation in Ecosystem-based Management', *Environmental Impact Assessment Review* 16, 71-102.

Endter-Wada, J. et al. (1998), 'A Framework for Understanding Social Science Contributions to Ecosystem Management', *Ecological Applications* 8:3, 891-904.

Grumbine, E. (1994), 'What Is Ecosystem Management?' *Conservation Biology* 8:1, 27-38.

Holling, C. (ed.) (1978), *Adaptive Environmental Assessment and Management* (New York: John Wiley & Sons).

Holling, C. (1995), 'What Barriers? What Bridges?' in L. Gunderson et al. (eds), *Barriers and Bridges to the Renewal of Ecosystems and Institutions* (New York: Columbia University Press), pp. 3-34.

Holling, C. (1996), 'Surprise for Science, Resilience for Ecosystems, and Incentives for People', *Ecological Applications* 6:3, 733-735.

Innes, J. (1996), 'Planning Through Consensus Building: A New View of the Comprehensive Planning Ideal', *Journal of American Planning Association* 62, 460.

Kessler, W. and Salwasser, H. (1995), 'Natural Resource Agencies: Transforming from Within' in R.L. Knight and S.F. Bates (eds), *A New Century for Resource Managers* (Covelo, CA: Island Press).

Lee, K. (1993), *Compass and Gyroscope: Integrating Science and Politics for the Environment* (Washington, DC: Island Press).

Lee, R. (1992), 'Ecologically Effective Social Organization as a Requirement for Sustaining Watershed Ecosystems', in R.J. Naiman (ed.), *Watershed Management: Balancing Sustainability and Environmental Change* (New York: Springer-Verlag).

Lessard, G. (1998), 'An Adaptive Approach to Planning and Decision-Making', *Landscape and Urban Planning* 40, 81-87.

Lowry, K. et al. (1997), 'Participating the Public: Group Processes, Politics, and Planning', *Journal of Planning Education and Research* 16, 177-187.

McGinnis, M. et al. (1999), 'Bioregional Conflict Resolution: Rebuilding Community in Watershed Planning and Organizing', *Environmental Management* 24:1, 1-12.

Ostrom, E. (1990), *Governing the Commons: The Evolution of Institutions for Collective Action* (New York: Cambridge University Press).

Power, T. (1996), *Lost Landscapes and Failed Economies: The Search for a Value of Place* (Washington, DC: Island Press).

Randolph, J. and Bauer, M. (1999), 'Improving Environmental Decision-making Through Collaborative Methods', *Policy Studies Review* 16, 169-191.

Westley, F. (1995), 'Governing Design: The Management of Social Systems and Ecosystems Management', in L. Gunderson et al. (eds), *Barriers and Bridges to the Renewal of Ecosystems and Institutions* (New York: Columbia University Press), pp. 391-427.

Wheatley, M. (1992), *Leadership and the New Science* (San Francisco: Berrett-Koehler Publishers, Inc).

Williams, D. and Stewart, S. (1998), 'Sense of Place: An Elusive Concept that is Finding a Home in Ecosystem Management', *Journal of Forestry* 96:5, 18-23.

Wondolleck, J. and Yaffee, S. (2000), *Making Collaboration Work: Lessons from Innovation in Natural Resource Management* (Washington, DC: Island Press).

Yaffee, S. (1996), 'Ecosystem Management in Practice', *Ecological Applications* 6:3, 724-727.

Yaffee, S. and Wondolleck, J. (1997), 'Building Bridges across Agency Boundaries', in Kathryn A. Kohm and Jerry F. Franklin (eds), *Creating A Forestry for the 21st Century* (Washington, DC: Island Press).

Chapter 4

Baer, W. (1997), 'General Plan Evaluation Criteria: An Approach to Making Better Plans', *Journal of the American Planning Association* 63:3, 329-344.

Beatley, T. (2000), 'Preserving Biodiversity: Challenges for Planners', *Journal of the American Planning Association* 66:1, 5-20.

Berke, P. and Manta, M. (2000), 'Are We Planning for Sustainable Development?' *Journal of the American Planning Association* 66:1, 21-33.

Berke, P. and French, S. (1994), 'The Influence of State Planning Mandates on Local Plan Quality', *Journal of Planning Education and Research* 13:4, 237-250.

Berke, P. et al. (1996), 'Enhancing Plan Quality: Evaluating the Role of State Planning Mandates for Natural Hazard Mitigation', *Journal of Environmental Planning and Management* 39, 79-96.

Berke, P. et al. (1998), *Do Co-operative Environmental Management Mandates Produce Good Plans?: The New Zealand Experience* (Chapel Hill, NC: DCRP).

Brody, S. (2003a), 'Implementing the Principles of Ecosystem Management Through Land Use Planning', *Population and Environment* 24:6, 511-520.

Brody, S. (2003b), 'Examining The Effects of Biodiversity on the Ability of Local Plans to Manage Ecological Systems', *Journal of Environmental Planning and Management* 46:6, 733-754.

Brody, S. et al. (2003a), 'Evaluating Ecosystem Management Capabilities at the Local Level in Florida: A Policy Gap Analysis Using Geographic Information Systems', *Environmental Management* 32:6, 661-681.

Brody, S. et al. (2004), 'Measuring the Collective Planning Capabilities of Local Jurisdictions To Manage Ecological Systems in Southern Florida', *Landscape and Urban Planning* 69:1, 33-50.

Brody, S. and Highfield. W. (2005), 'Does Planning Work? Testing the Implementation of Local Environmental Planning in Florida', *Journal of the American Planning Association* 71:2, 159-175.

Burby, R. and May, P. (1998), 'Intergovernmental Environmental Planning: Addressing the Commitment Conundrum', *Journal of Environmental Planning and Management* 41:1, 95-110.

Burby, R. and Dalton, L. (1994), 'Plans Can Matter! The Role of Land Use Plans and State Planning Mandates in Limiting the Development of Hazardous Areas', *Public Administration Review* 54:3, 229-237.

Burby, R. and May, P., with Berke, P., Dalton, L., French, S., and Kaiser, E. (1997), *Making Governments Plan: State Experiments in Managing Land Use* (Baltimore: Johns Hopkins University Press).

Chapin, S. and Kaiser, E. (1979), *Urban Land Use Planning* (Illinois: University of Illinois Press).

Daniels, S. et al. (1996), 'Decision-making and Ecosystem-based Management: Applying the Vroom-Yetton Model to Public Participation Strategy' *Environmental Impact Assessment Review* 16, 13-30.

Daniels, S. and Walker, G. (1996), 'Collaborative Learning: Improving Public Deliberation in Ecosystem-based Management', *Environmental Impact Assessment Review* 16, 71-102.

Duerksen, C. et al. (1997), *Habitat Protection Planning: Where the Wild Things Are*, APA Planning Advisory Report No. 470/471.

Godschalk, D. et al. (1999), *Natural Hazard Mitigation* (Washington, DC: Island Press).

Grumbine, E. (1994), 'What Is Ecosystem Management?' *Conservation Biology* 8:1, 27-38.

Hoch, C. (1998), 'Evaluating Plan Pragmatically', Paper presented at the 40th Annual Conference, Association of Collegiate Schools of Planning, Pasadena, CA, November 5.

Kaiser, E. and Godschalk, D. (2000), 'Development Planning', in L. Dalton (ed.), *The Practice of Local Government Planning*, 3rd edition (Washington, DC: International City/County Management Association).

Kaiser, E. et al. (1995), *Urban Land Use Planning*, 4th edition (Urbana: University of Illinois Press).

Kirklin, J. (1995), 'Protecting Species and Ecosystems Within Planning Processes', *Environmental Planning* 12:4, 6-13.

Light, S. (1995), 'The Everglades: Evolution of Management in a Turbulent Ecosystem', in L. Gunderson et al. (eds), *Barriers and Bridges to the Renewal of Ecosystems and Institutions* (New York: Columbia University Press), pp. 3-34.

Randolph, J. and Bauer, M. (1999), 'Improving Environmental Decision-making Through Collaborative Methods', *Policy Studies Review* 16, 169-191.

Talen, E. (1996), 'Do Plans Get Implemented? A Review of Evaluation in Planning', *Journal of Planning Literature* 10:3, 248-59.

Vogt, J. et al. (1997), *Ecosystems* (New York: Springer-Verlag).

Wondolleck, J. and Yaffee, S. (2000), *Making Collaboration Work: Lessons from Innovation in Natural Resource Management* (Washington, DC: Island Press).

Chapter 5

Anselin, L. (1995), 'Local Indicators of Spatial Association – LISA', *Geographical Analysis* 27:2, 93-115.

Asthon, R. and Asthon, P. (1988), *Handbook of Reptiles and Amphibians of Florida. Part Three: The Amphibians* (Miami, FL: Windward).

Beatley, T. (2000), 'Preserving Biodiversity: Challenges for Planners', *Journal of the American Planning Association* 66:1, 5-20.

Berke, P. and French, S. (1994), 'The Influence of State Planning Mandates on Local Plan Quality', *Journal of Planning Education and Research* 13:4, 237-250.

Berke, P. et al. (1996), 'Enhancing Plan Quality: Evaluating the Role of State Planning Mandates for Natural Hazard Mitigation', *Journal of Environmental Planning and Management* 39, 79-96.

Berke, P. et al. (1998), *Do Co-operative Environmental Management Mandates Produce Good Plans?: The New Zealand Experience* (Chapel Hill, NC: DCRP).

Brody, S. (2003), 'Implementing the Principles of Ecosystem Management Through Land Use Planning', *Population and Environment* 24:6, 511-520.

Brody, S.D., Carassco, V. and Highfield, W. (2003), 'Evaluating Ecosystem Management Capabilities at the Local Level in Florida: Identifying Policy Gaps Using Geographic Information Systems', *Environmental Management* 32:6, 661-681.

Cox, J. et al. (1994), *Closing the Gaps in Florida's Wildlife Habitat Conservation System* (Tallahassee: Florida Game and Fresh Water Fish Commission).

Duerksen, C. et al. (1997), *Habitat Protection Planning: Where the Wild Things Are*, APA Planning Advisory Report No. 470/471.

Ferriter, A. (ed.) (1997), *Brazilian Pepper Management Plan for Florida* (West Palm Beach: Florida EPPC).

Foster, M. and Humphrey, S. (1995), 'Use of Underpasses by Florida Panthers and Other Wildlife', *Wildlife Society Bulletin* 23, 95-100.

Fuller, P. and Benson, A. (1999), 'Nonindigenous Species Introduced into South Florida', presented at the South Florida Restoration Science Forum, May 17-19, Boca Raton, FL.

Godschalk, D. et al. (1999), *Natural Hazard Mitigation* (Washington, DC: Island Press).

Lee, K. (1993), *Compass and Gyroscope: Integrating Science and Politics for the Environment* (Washington, DC: Island Press).

Mahendra, K. et al. (1995), 'Specific Competitive Inhibitor of Secreted Phospholase A_2 from Berries of *Schinus terebinthifolius*', *Phytochemistry* 39, 537-47.

Miles, M. and Huberman, A. (1984), *Qualitative Data Analysis: A Sourcebook of New Methods* (Newbury Park, CA: Sage).

Noss, R., and Cooperrider, A. (1994), *Saving Nature's Legacy: Protecting and Restoring Biodiversity* (Washington, DC: Island Press).

Peck, S. (1998), *Planning for Biodiversity: Issues and Examples* (Washington, DC: Island Press).

Smith, D.S. (1993), 'Greenway Case Studies', in D.S. Smith and P.C. Hellmund (eds), *Ecology of Greenways: Design and Function of Linear Conservation Areas* (Ann Arbor: University of Michigan Press), pp. 161-208.

Chapter 6

Birkland, T. (1997), *After Disaster: Agenda Setting, Public Policy, and Focusing Events* (Washington, DC: Georgetown University Press).
Burby, R. and French, S. (1981), 'Coping With Floods: The Land Use Management Paradox', *Journal of the American Planning Association* 47:3, 289-300.
Davis, F. et al. (1990), 'An Information Systems Approach to the Preservation of Biological Diversity', *International Journal of Geographical Information Systems* 4:1, 55-78.
Dramstad, W. et al. (1996), *Landscape Ecology Principles in Landscape Architecture and Land-use Planning* (Washington, DC: Island Press).
Duerksen, C. et al. (1997), *Habitat Protection Planning: Where the Wild Things Are*, APA Planning Advisory Report No. 470/471.
Edwards, T. et al. (1993), 'Gap Analysis: A Geographic for Assessing National Biological Diversity', *Natural Resources and Environmental Issues* 2, 65-71.
Forman, R. and Gordon, M. (1986), *Landscape Ecology* (New York: Wiley).
Forman, R. (1995a), *Land Mosaics: The Ecology of Landscapes and Regions* (Cambridge: Cambridge University Press).
Forman, R. (1995b), 'Some General Principles of Landscape and Regional Ecology', *Landscape Ecology* 10, 133-142.
Grumbine, E. (1990), 'Protecting Biological Diversity through the Greater Ecosystem Concept', *Natural Areas Journal* 10:3, 114-120.
Grumbine, E. (1994), 'What is Ecosystem Management?' *Conservation Biology* 8:1, 27-38.
Haeubner, R. (1998), 'Ecosystem Management and Environmental Policy in the United States: Open Window or Closed Door?' *Landscape and Urban Planning* 40, 221-233.
Harris, L. (1984), *The Fragmented Forest: Island Biogeography Theory and the Preservation of Biotic Diversity* (Chicago: University of Chicago Press).
Jongman, R. et al. (2004), 'European Ecological Networks and Greenways', *Landscape and Urban Planning* 68:2-3, 305-319.
Kingdon, J. (1984), *Agendas, Alternatives and Public Policy* (Boston: Brown & Co.).
Kirklin, J. (1995), 'Protecting Species and Ecosystems within Planning Processes', *Environmental Planning* 12:4, 6-13.
Lein, J. (2003), *Integrated Environmental Planning* (London: Blackwell Science, Ltd.).
Lindell, M. and Perry, R. (1999), 'Household Adjustment to Earthquake Hazard: A Review of Research', *Environment and Behavior* 32:4, 590-630.
Lindell, M. and Prater, C. (2000), 'Household Adoption of Seismic Hazard Adjustments: A Comparison of Residents in Two States', *International Journal of Mass Emergencies and Disasters* 18:2, 317-338.
McCormick, F. (1999), 'Principles of Ecosystem Management and Sustainable Development', in J. Peine (ed.), *Ecosystem Management for Sustainability* (New York: Lewis Publishers).

Murry, S. et al. (1999), 'No-take Reserve Networks: Sustaining Fishery Populations and Marine Ecosystems', *Fisheries* 24:11, 11-24.

McNeely, J. (1992), 'The Biodiversity Crisis: Challenges for Research and Management', in O.T. Sandlund et al. (eds), *Conservation of Biodiversity for Sustainable Development* (Norway: Scandinavian University Press).

Noss, R. (1983), 'A Regional Landscape Approach to Maintain Diversity', *BioScience* 33, 700-706.

Noss, R. (1991), 'Landscape Connectivity: Different Functions at Different Scales', in W.E. Hudson (ed.), *Landscape Linkages and Biodiversity* (Washington, DC: Island Press).

Noss, R., and Cooperrider, A. (1994), *Saving Nature's Legacy: Protecting and Restoring Biodiversity* (Washington, DC: Island Press).

Noss, R., and Scott, M. (1997), 'Ecosystem Protection and Restoration: The Core of Ecosystem Management', in M. Boyce and A. Hanley (eds), *Ecosystem Management: Applications for Sustainable Forest and Wildlife Resources* (New Haven: Yale University Press), pp. 239-264.

Peck, S. (1998), *Planning for Biodiversity: Issues and Examples* (Washington, DC: Island Press).

Ray, C. (1996), 'Biodiversity is Biogeography: Implications for Conservation', *Oceanography* 9:1, 50-59.

Ruth, S. (1990), 'Risk Identification Techniques for Land Managers – An Analysis of Current Legal Standards', *Urban Wildlife Manager's Notebook – 18* 11:3, 1-4.

Scott, J. et al. (1991), 'Gap Analysis: Assessing Protection Needs', in W. Hudson (ed.), *Landscape Linkages and Biodiversity: A Strategy for Survival* (Seattle, WA: Island Press).

Scott, J. et al. (1993), 'Gap Analysis: A Geographical Approach to Protection of Biological Diversity', *Wildlife Monographs* 123, 1-41.

Shafer, C. (1995), 'Values and Shortcomings of Small Reserves', *BioScience* 45:2, 80-88.

Slocombe, S. (1998), 'Defining Goals and Criteria for Ecosystem-based Management', *Environmental Management* 22:4, 483-493.

Soule, M. (1991), 'Land Use Planning and Wildlife Maintenance: Guidelines for Conserving Wildlife in an Urban Landscape' *Journal of the American Planning Association* 57:3, 313-323.

Turner, R. et al. (1986), *Waiting for Disaster: Earthquake Watch in California* (Berkeley, CA: University of California Press).

Van Langevelde, F. (1994), 'Conceptual Integration of Landscape Planning and Landscape Ecology, with a Focus on the Netherlands', in E.A. Cook and H.N. van Lier (eds), *Landscape Planning and Ecological Networks* (Amsterdam: Elsevier Science).

Van Lier, H.N. and Cook, E. (1994), 'Ecological Networks: a Conspectus', in E.A. Cook and H.N. van Lier (eds), *Landscape Planning and Ecological Networks* (Amsterdam: Elsevier Science).

Vogt, J. et al. (1997), *Ecosystems* (New York: Springer-Verlag).

Wondolleck, J. and Yaffee, S. (2000), *Making Collaboration Work: Lessons from Innovation in Natural Resource Management* (Washington, DC: Island Press).

Chapter 7

Berke, P. et al. (1996), 'Enhancing Plan Quality: Evaluating the Role of State Planning Mandates for Natural Hazard Mitigation', *Journal of Environmental Planning and Management* 39, 79-96.

Berke, P. et al. (1998), *Do Co-operative Environmental Management Mandates Produce Good Plans?: The New Zealand Experience* (Chapel Hill, NC: DCRP).

Brody, S. (2001), *Pinellas County: The Role of Focused Participation in the Comprehensive Planning Process* (Chapel Hill, NC: Center for Urban and Regional Studies).

Brody, S. (2003), 'Examining the Effects of Biodiversity on the Ability of Local Plans to Manage Ecological Systems', *Journal of Environmental Planning and Management* 46:6, 733-754.

Brody, S.D., Highfield, W. and Carrasco, V. (2004), 'Measuring the Collective Capabilities of Local Jurisdictions to Manage Ecosystems in Southern Florida', *Journal of Landscape and Urban Planning* 69:1, 33-50.

Burby, R. and May, P. (1998), 'Intergovernmental Environmental Planning: Addressing the Commitment Conundrum', *Journal of Environmental Planning and Management* 41:1, 95-110.

Cox, J. et al. (1994), *Closing the Gaps in Florida's Wildlife Habitat Conservation System* (Tallahassee: Florida Game and Fresh Water Fish Commission).

Fransson, N. and Garling, T. (1999), 'Environmental Concern: Conceptual Definitions, Measurements, Methods, and Research Findings', *Journal of Environmental Psychology* 19, 369-382.

Guagano, A. and Markee, N. (1995), 'Regional Differences in the Socio-demographic Determinants of Environmental Concern', *Population and Environment* 17:2, 135-149.

Haeubner, R. (1998), 'Ecosystem Management and Environmental Policy in the United States: Open Window or Closed Door?' *Landscape and Urban Planning* 40, 221-233.

Howell, S. and Laska, S. (1992), 'The Changing Face of the Environmental Coalition: A Research Note', *Environment and Behavior* 24,134-144.

Raudsepp, M. (2001), 'Some Socio-demographic and Socio-psychological Predictors of Environmentalism', *Trames* 5:55/50, 355-367.

Scott, D. and Willits, F. (1994), 'Environmental Attitudes and Behavior', *Environment and Behavior* 26:2, 239-261.

Van Liere, K. and Dunlap, R. (1981), 'Environmental Concern – Does it Make a Difference How it's Measured?' *Environment and Behavior* 13, 651-676.

Williams, J. et al. (eds) (1997), *Watershed Restoration: Principles and Practices* (Bethesda, MD: American Fisheries Society).

Chapter 8

Alterman, R. et al. (1984), The Impact of Public Participation on Planning', *Town Planning Review* 55:2, 177-196.

Arnstein, S. (1969), 'A Ladder of Citizen Participation', *Journal of the American Planning Association* 35:4, 216-224.

Beatley, T. et al. (1994), 'Representation in Comprehensive Planning', *Journal of the American Planning Association* 60:2, 185-196.

Beierle, T. and Konisky, D. (2001), 'What are We Gaining from Stakeholder Involvement? Observation from Environmental Planning in the Great Lakes', *Journal of Environmental Planning C: Government and Policy* 19:4, 515-527.

Beyer, D. et al. (1997), 'Ecosystem Management in the Eastern Upper Peninsula of Michigan: A Case History', *Landscape and Urban Planning* 38, 199-211.

Bingham, G. (1986), *Resolving Environmental Disputes: A Decade of Experience.* (Washington, DC: Conservation Foundation).

Brody, S.D. (2003), 'Evaluating the Role of Resource-Based Industries in Ecosystem Approaches to Management' *Journal of Society and Natural Resources* 16:7, 625-641.

Brody, S. (2001), *Public Participation in the City of Fort Lauderdale Comprehensive Plan: A Constituency Model of Planning Making* (Chapel Hill, NC: Center for Urban and Regional Studies).

Burke, E. (1979), *A Participatory Approach to Urban Planning* (New York: Human Science Press).

Carpenter, S. and Kennedy, J. (1988), *Managing Public Disputes: A Practical Guide to Handling Conflict and Reaching Agreement* (San Francisco: Jossey-Bass).

Creighton, J. (1992), *Involving Citizens in Decision Making: A Guidebook* (Washington, DC: Program for Community Problem Solving).

Crowfoot, J. and Wondolleck, J. (1990), *Environmental Disputes: Community Involvement in Conflict Resolution* (Washington, DC: Island Press).

Day, D. (1997), 'Citizen Participation in the Planning Process: An Essentially Contested Concept?' *Journal of Planning Literature* 11:3, 421-434.

Duane, T. (1997), 'Community Participation in Ecosystem Management', *Ecology Law Quarterly* 24:4, 771-797.

Duram, L. and Brown, K. (1999), 'Assessing Public Participation in US Watershed Planning Initiatives', *Society and Natural Resources* 12:5, 455-467.

Fainstein, N. and Fainstein, S. (1985), 'Citizen Participation in Local Government', in D.R. Judd (ed.), *Public Policy Across States and Communities* (Greenwich, CT: Jai Press), pp. 223-238.

Godschalk, D. et al. (1994), *Pulling Together: A Planning and Development Consensus Building Manual* (Washington, DC: Urban Land Institute).

Gray, B. (1989), *Collaborating: Finding Common Ground for Multiparty Problems* (San Francisco: Jossey-Bass).

Grumbine, E. (1994), 'What Is Ecosystem Management?' *Conservation Biology* 8:1, 27-38.

Hoffman, A. et al. (1997) 'Balancing Business Interests and Endangered Species Protection', *Sloan Management Review* 39:1, 59-73.

Howell, R. et al. (1987), *Designing a Citizen Involvement Program: A Guidebook for Involving Citizens in the Resolution of Environmental Issues* (Corvallis, OR: Western Rural Development Center, Oregon State University).

Innes, J. (1996), 'Planning Through Consensus Building: A New View of the Comprehensive Planning Ideal', *Journal of American Planning Association* 62, 460.

Jones, S. (1994), 'Ecosystem Management on NIPF: A Mandate for Cooperative Education', *Journal of Forestry* 92:8, 14-15.

Kaza, S. (1998), 'Community Involvement in Marine Protected Areas', *Oceanus* 31:1, 75-88.

Kennedy, D. et al. (2000), *The New Watershed Source Book* (Boulder, CO: Natural Resources Law Center, University of Colorado School of Law).

Machlis, G. (1999), 'New Forestry, Neopolitics, and Voodoo Economies: Research Needs for Biodiversity Management', in J. Aley et al. (eds), *Ecosystem Management: Adaptive Strategies for Natural Resources Organizations in the Twenty-First Century* (Philadelphia: Taylor & Francis).

MacKenzie, S. (1996), *Integrated Resource Planning and Management: The Ecosystem Approach in the Great Lakes Basin* (Washington, DC: Island Press).

McCool, S. and Guthrie, K. (2001), 'Mapping the Dimensions of Successful Public Participation in Messy Natural Resources Management Situations', *Society and Natural Resources* 14:4, 309-323.

Moore, N. (1995), *Participation Tools for Better Land-Use Planning: Techniques and Case Studies* (Sacramento, CA: Local Government Commission).

O'Connell, M. (1996), 'Managing Biodiversity on Private Lands', in R. Szaro and D. Johnston (eds), *Biodiversity in Managed Landscapes: Theory and Practice* (Oxford: Oxford University Press).

Vogt, J. et al. (1997), *Ecosystems* (New York: Springer-Verlag).

Westley, F. (1995), 'Governing Design: The Management of Social Systems and Ecosystems Management', in L. Gunderson et al. (eds), *Barriers and Bridges to the Renewal of Ecosystems and Institutions* (New York: Columbia University Press), pp. 391-427.

Wondolleck, J. and Yaffee, S. (2000), *Making Collaboration Work: Lessons from Innovation in Natural Resource Management* (Washington, DC: Island Press).

Yaffee, S. et al. (1996), *Ecosystem Management in the United States: An Assessment of Current Experience* (Washington, DC: Island Press).

Yaffee, S. and Wondolleck, J. (1997), 'Building Bridges across Agency Boundaries', in K. Kohm and J. Franklin (eds), *Creating A Forestry for the 21st Century* (Washington, DC: Island Press).

Chapter 9

Beatley, T. et al. (1994), 'Representation in Comprehensive Planning', *Journal of the American Planning Association* 60:2, 185-196.

Beierle, T. and Konisky, D. (2001), 'What are we Gaining from Stakeholder Involvement? Observation from Environmental Planning in the Great Lakes', *Journal of Environmental Planning C: Government and Policy* 19:4, 515-527.

Brody, S. (2001a), *The City of Sarasota, FL 1998 Comprehensive Plan: The Role of Communicative Culture and Informal Public Participation in Plan Making* (Chapel Hill, NC: Center for Urban and Regional Studies).
Brody, S. (2001b), *Public Participation in the City of Fort Lauderdale Comprehensive Plan: A Constituency Model of Planning Making* (Chapel Hill, NC: Center for Urban and Regional Studies).
Brody, S. (2001c), *Pinellas County: The Role of Focused Participation in the Comprehensive Planning Process* (Chapel Hill, NC: Center for Urban and Regional Studies).
Brody, S. (2003), 'Examining the Impacts of Stakeholder Participation in Watershed Approaches to Planning', *Journal of Planning Education and Research* 22:4, 107-119.
Brody, S. et al. (2003), 'Mandating Citizen Participation in Plan-Making: Six Strategic Choices', *Journal of American Planning Association* 69:3, 245-265.
Crowfoot, J. and Wondolleck, J. (1990), *Environmental Disputes: Community Involvement in Conflict Resolution* (Washington, DC: Island Press).
Susskind, L. et al. (eds) (1999), *Consensus Building Handbook: A Comprehensive Guide to Reaching Agreement* (California: Sage Publications).

Chapter 10

Alexander, E. and Faludi, A. (1989), 'Planning and Plan Implementation: Notes on Evaluation Criteria', *Environment and Planning B: Planning and Design* 16:2, 127-140.
Alonso, W. (1964), *Location and Land Use: Toward a General Theory of Land Rent* (Cambridge MA: Harvard University Press).
Alterman, R. and Hill, M. (1978), 'Implementation of Urban Land Use Plans', *Journal of the American Institute of Planners* 44:3, 274-285.
Baer, W. (1997), 'General Plan Evaluation Criteria: An Approach to Making Better Plans', *Journal of the American Planning Association* 63:3, 329-344.
Bengston, D. et al. (2004), 'Public Policies for Managing Urban Growth and Protecting Open Space: Policy Instruments and Lessons Learned in the United States', *Landscape and Urban Planning* 69:2-3, 271-286.
Berke, P. and French, S. (1994), 'The Influence of State Planning Mandates on Local Plan Quality', *Journal of Planning Education and Research* 13:4, 237-250.
Berke, P. et al. (2006), 'What Makes Plan Implementation Successful? An Evaluation of Local Plans and Implementation Practices in New Zealand', *Environment and Planning B: Planning and Design* 33:4, 581-600.
Brody, S. (2003), 'Examining the Impacts of Stakeholder Participation in Watershed Approaches to Planning', *Journal of Planning Education and Research* 22:4, 107-119.
Brody, S. (2003b), 'Implementing the Principles of Ecosystem Management Through Land Use Planning', *Population and Environment* 24:6, 511-520.

Brody, S. and Highfield. W. (2005), 'Does Planning Work? Testing the Implementation of Local Environmental Planning in Florida', *Journal of the American Planning Association* 71:2, 159-175.

Brueckner, J. (2000), 'Urban Sprawl: Diagnosis and Remedies', *International Regional Science Review* 23:2, 160-171.

Bryson, J. (1991), 'There is No Substitute for an Empirical Defense of Planning and Planners', *Journal of Planning Education and Research* 10:2, 164-165.

Burby, R. and May, P. (1998), 'Intergovernmental Environmental Planning: Addressing the Commitment Conundrum', *Journal of Environmental Planning and Management* 41:1, 95-110.

Burby, R. et al. (1997), *Making Governments Plan: State Experiments in Managing Land Use* (Baltimore: Johns Hopkins University Press).

Burby, R. (2003), 'Making Plans that Matter: Citizen Involvement and Government Action', *Journal of the American Planning Association* 69:1, 33-49.

Calkins, H. (1979), 'The Planning Monitor: An Accountability Theory of Plan Evaluation', *Environment and Planning A* 11:7, 745-758.

Carruthers, J. (2002), 'The Impacts of State Growth Management Programmes: A Comparative Analysis', *Urban Studies* 39:11, 1959-1982.

Carruthers, J. (2003), 'Growth at the Fringe: The Influence of Political Fragmentation in United States Metropolitan Areas', *Papers in Regional Science* 82:4, 475-499.

Carruthers, J.I. and Ulfarsson, G.F. (2002), 'Fragmentation and Sprawl: Evidence from Interregional Analysis', *Growth and Change* 33:3, 312-340.

Clawson, M. (1971), *Suburban Land Conversion in the United States: An Economic and Governmental Process* (Baltimore: Johns Hopkins University Press).

Daniels, T. (1999), *When City and Country Collide: Managing Growth in the Metropolitan Fringe* (Washington, DC: Island Press).

Driessen, P. (1997), 'Performance and Implementing Institutions in Rural Land Development', *Environment and Planning B: Planning and Design* 24:6, 859-869.

Faludi, A. (2000), 'The Performance of Spatial Planning', *Planning Practice and Research* 15:4, 299-318.

Gillham, O. (2002), *The Limitless City: A Primer on the Urban Sprawl Debate* (Washington, DC: Island Press).

Heimlich, R. and Anderson, W. (2001), *Development at the Urban Fringe and Beyond: Impacts on Agriculture and Rural Land*, Economic Research Service, US Department of Agriculture, Agricultural Economic Report No. 803.

Knapp, G. (1985), 'The Price Effects of an Urban Growth Boundary in Metropolitan Portland, Oregon', *Land Economics* 61:1, 26-35.

Mastrop, H. and Faludi, A. (1997), 'Evaluation of Strategic Plans: The Performance Principle', *Environment and Planning B: Planning and Design*, 24:6, 815-832.

Mattson, G. (2003), *Small Towns, Sprawl and the Politics of Policy Choices: The Florida Experience* (Latham, MD: University Press of America).

Murtagh, B. (1998), 'Evaluating the Community Impacts of Urban Policy', *Planning Practice and Research* 13:2, 129-138.

Pendall, R. (1999), 'Do Land Use Controls Cause Sprawl?' *Environment and Planning B* 26:4, 555-571.

Seasons, M. (2003), 'Monitoring and Evaluation in Municipal Planning', *Journal of the American Planning Association* 69:4, 430-440.

Shen, Q. (1996), 'Spatial Impacts of Locally Enacted Growth Controls: The San Francisco Bay Area in the 1980s', *Environment and Planning B: Planning and Design* 23:1, 61-91.

Talen, E. (1996a), 'Do Plans Get Implemented? A Review of Evaluation in Planning', *Journal of Planning Literature* 10:3, 248-259.

Talen, E. (1996b), 'After the Plans: Methods to Evaluate the Implementation Success of Plans', *Journal of Planning Education and Research* 16:2, 79-91.

Talen, E. (1997), 'Success, Failure, and Conformance: An Alternative Approach to Planning Evaluation', *Environment and Planning B: Planning and Design* 24:4, 573-587.

Wildavsky, A. (1973), 'If Planning is Everything, Maybe it's Nothing', *Policy Sciences* 4:2, 127-153.

Zhang, T. (2001), 'Community Features and Urban Sprawl: The Case of the Chicago Metropolitan Region', *Land Use Policy* 18:3, 221-232.

Chapter 11

Anselin, L. (1995), 'Local Indicators of Spatial Association – LISA', *Geographical Analysis* 27:2, 93-115.

Beatley, T. (2000), 'Preserving Biodiversity: Challenges for Planners', *Journal of the American Planning Association* 66:1, 5-20.

Berke, P. and French, S. (1994), 'The Influence of State Planning Mandates on Local Plan Quality', *Journal of Planning Education and Research* 13:4, 237-250.

Berke, P. et al. (2006), 'What Makes Plan Implementation Successful? An Evaluation of Local Plans and Implementation Practices in New Zealand', *Environment and Planning B: Planning and Design* 33:4, 581-600.

Brody, S. (2003), 'Examining the Impacts of Stakeholder Participation in Watershed Approaches to Planning', *Journal of Planning Education and Research* 22:4, 107-119.

Brody, S. et al. (2003), 'Mandating Citizen Participation In Plan-Making: Six Strategic Choices', *Journal of American Planning Association* 69:3, 245-265.

Brody, S. and Highfield, W. (2005), 'Does Planning Work? Testing the Implementation of Local Environmental Planning in Florida', *Journal of the American Planning Association* 71:2, 159-175.

Brody, S. et al. (2006a), 'Measuring the Adoption of Local Sprawl Reduction Planning Policies in Florida', *Journal of Planning Education and Research* 25:3, 294-310.

Brody, S. et al. (2006b), 'Planning at the Urban Fringe: An Examination of the Factors Influencing Nonconforming Development Patterns in Southern Florida', *Environment and Planning B* 33:1, 75-96.

Burby, R. (2003), 'Making Plans that Matter: Citizen Involvement and Government Action', *Journal of the American Planning Association* 69:1, 33-49.

Duerksen, C. et al. (1997), *Habitat Protection Planning: Where the Wild Things Are*, APA Planning Advisory Report No. 470/471.

Mattson, G. (2003), *Small Towns, Sprawl and the Politics of Policy Choices: The Florida Experience* (Latham, MD: University Press of America).

Williams, J. et al. (eds) (1997), *Watershed Restoration: Principles and Practices* (Bethesda, MD: American Fisheries Society).

Chapter 12

Beatley, T. and Manning, K. (1997), *Ecology of Place: Planning for Environment, Economy and Community* (Washington, DC: Island Press).

Beatley, T. (2000), 'Preserving Biodiversity: Challenges for Planners', *Journal of the American Planning Association* 66:1, 5-20.

Brody, S. et al. (2006), 'Examining Motivations for Resource-Based Industry to Participate in Collaborative Ecosystem Management Initiatives', *Forest Policy and Economics* 8:2, 123-134.

Duerksen, C. et al. (1997), *Habitat Protection Planning: Where the Wild Things Are*, APA Planning Advisory Report No. 470/471.

Endter-Wada, J. et al. (1998), 'A Framework for Understanding Social Science Contributions to Ecosystem Management', *Ecological Applications* 8:3, 891-904.

Hoffman, A. (2000), *Competitive Environmental Strategy: A Guide to the Changing Business Landscape* (Washington, DC: Island Press).

Holling, C. (1995), 'What Barriers? What Bridges?' in L. Gunderson et al. (eds), *Barriers and Bridges to the Renewal of Ecosystems and Institutions* (New York: Columbia University Press), pp. 3-34.

Pittman, C. (2003), 'Florida's Land Czar', *Planning* 69: 4-9.

Index

References such as "178-9" indicate (not necessarily continuous) discussion of a topic across a range of pages, whilst "97t7.2" refers to Table 7.2 on page 97 and "63f5.3" to Figure 5.3 on page 63. Wherever possible in the case of topics with many references, these have either been divided into sub-topics or the most significant discussions of the topic are indicated by page numbers in bold. Because the entire volume is about ecosystems, planning and Florida, the use of these terms (and certain others which occur constantly throughout the book) as entry points has been minimized. Information will be found under the corresponding detailed topics.